ASF-5488

PROTECTING THE OZONE LAYER:

Lessons, Models, and Prospects

PROTECTING THE OZONE LAYER:

Lessons, Models, and Prospects

edited by

Philippe G. Le Prestre
Université du Québec à Montréal

John D. Reid
Environment Canada

E. Thomas Morehouse, Jr.
Institute for Defense Analyses, U.S.A.

KLUWER ACADEMIC PUBLISHERS
Boston / Dordrecht / London

Distributors for North, Central and South America:
Kluwer Academic Publishers
101 Philip Drive
Assinippi Park
Norwell, Massachusetts 02061 USA
Telephone (781) 871-6600
Fax (781) 871-6528
E-Mail <kluwer@wkap.com>

Distributors for all other countries:
Kluwer Academic Publishers Group
Distribution Centre
Post Office Box 322
3300 AH Dordrecht, THE NETHERLANDS
Telephone 31 78 6392 392
Fax 31 78 6546 474
E-Mail <orderdept@wkap.nl>

 Electronic Services <http://www.wkap.nl>

Library of Congress Cataloging-in-Publication Data

Protecting the ozone layer : lessons, models, and
 prospects / edited by Philippe G. Le Prestre, John D.
 Reid, E. Thomas Morehouse.
      p.   cm.
   Includes bibliographical references and index.
   ISBN 0-7923-8245-5
   1. Ozone layer--Environmental aspects.   I. Le
Prestre, Philippe G.   II. Reid, John D.
III. Morehouse, E. Thomas
QC881.2.O9P76   1998   98-30071
363.738'75--dc21   CIP

Copyright © 1998 by Kluwer Academic Publishers.

All rights reserved. No part of this publication may be reproduced, stored in a retrieval system or transmitted in any form or by any means, mechanical, photocopying, recording, or otherwise, without the prior written permission of the publisher, Kluwer Academic Publishers, 101 Philip Drive, Assinippi Park, Norwell, Massachusetts 02061

Printed on acid-free paper.

Printed in the United States of America

Dedication

To Mostafa K. Tolba, for eighteen years Executive Director of the United Nations Environment Programme, whose vision and leadership united the international community in action to protect and restore the ozone layer.

Contents

DEDICATION v

LIST OF CONTRIBUTORS ix

ACKNOWLEDGEMENTS xv

LIST OF ACRONYMS xvii

INTRODUCTION
The Montreal Regime: A New Model for International Cooperation on Global Environmental Issues? 1
    Philippe G. Le Prestre, John D. Reid and E. Thomas Morehouse

THE SCIENTIST AND THE POLITICIAN
Stratospheric Ozone Chemistry 13
    Mario Molina
The Story of the Ozone Layer: Lessons Learned and Impacts on the Future 19
    Mostafa K. Tolba

THE ROLE OF SCIENCE
The Evolving UV Climate 29
    James B. Kerr
UV-B Effects on Aquatic Ecosystems 39
    Robert Worrest
Effects of UV-B on Plants and Terrestrial Ecosystems 43
    Manfred Tevini
Stratospheric Ozone Depletion and UV-Induced Immune Suppression 47
    Edward C. De Fabo
Effects on Skin and Eyes 55
    Jan C. van der Leun
UV INDEX: A Tool for Public Response 59
    Anne O'Toole
What Should Be Done in a Science Assessment 67
    Daniel Albritton
Commentary on Presentation by Daniel Albritton 75
    Gordon McBean

THE ROLE OF DIPLOMACY
The Montreal Protocol as a New Approach to Diplomacy 81
    Richard Elliot Benedick
The Montreal Protocol: A New Legal Model for Compliance Control 91
    Patrick Széll
The Use of Trade Measures in the Montreal Protocol 99
    Duncan Brack
The Montreal Regime: Sticks and Carrots 107
    Peter H. Sand

International Cooperation: An Example of Success *113*
    Juan Antonio Mateos
The Montreal Protocol: Whose Model? *117*
    Ashok Khosla
Comments on Ambassador Mateos' and Dr. Khosla's Remarks *123*
    Victor Buxton
The Montreal Protocol: The First Adaptive Global Environmental Regime? ... *127*
    Edward A. Parson
Comments: The Atmosphere as Global Commons *135*
    Marvin S. Soroos

THE ROLE OF TECHNOLOGY
Technology Assessment for the Montreal Protocol *143*
    Suely Machado Carvalho
TEAP Terms of Reference *149*
    Robert Van Slooten
Countries with Economies in Transition *153*
    László Dobó and Lambert Kuijpers
Importance of the TEAP in Technology Cooperation *163*
    Sally Rand and Lalitha Singh
Scientific Objectivity, Industrial Integrity and the TEAP Process *167*
    Lambert Kuijpers, Helen Tope, Jonathan Banks, Walter Brunner and Ashley
    Woodcock
Global Benefits and Costs of the Montreal Protocol *173*
    James Armstrong
Lessons From the CFC Phase-out in the United States *179*
    Elizabeth Cook
*Highlights of Ozone Protection Leadership by Industry in Developing
Countries* *191*
    Jorge Corona and José I. Pons
Competitive Advantage Through Corporate Environmental Leadership *195*
    Margaret G. Kerr
*Lessons from the Thailand Leadership Initiative, the Vietnam Corporate
Pledge, and the Global Semiconductor Agreement* *201*
    Yuichi Fujimoto
Champions of Ozone Layer Protection *207*
    Stephen O. Andersen
Closing Comments *213*
    Gary Taylor and E. Thomas Morehouse

ANNEX 1
Ozone Protection Chronology *217*

ANNEX 2
The Tenth Anniversary Colloquium Declaration *221*

INDEX *227*

List of Contributors

Albritton, Daniel
Dr. Albritton is Director of the Aeronomy Laboratory of the U.S. National Oceanic and Atmospheric Administration since 1986, and Co-Chair of the UNEP Scientific Assessment Panel for the Montreal Protocol.

Andersen, Stephen O.
Dr. Andersen is a Director within the Stratospheric Protection and the Atmospheric Pollution Prevention Divisions of the U.S. Environmental Protection Agency and Co-Chair of the UNEP Technology and Economic Assessment Panel (TEAP) for the Montreal Protocol.

Armstrong, James
Mr. Armstrong is chief of the Chemicals Control Division of the Environmental Protection Service of Environment Canada.

Banks, Jonathan
Dr. Banks is a research scientist at the Stored Grain Research Laboratory. He is the Co-Chair of the Methyl Bromide Technical Options Committee (MBTOC) and, in that capacity, a member of the Technology and Economic Assessment Panel (TEAP).

Benedick, Richard Elliot
Ambassador Benedick was the chief U.S. negotiator during the creation of the Montreal Protocol in 1987 and served as special advisor to the secretaries-general of both the 1992 United Nations Conference on Environment and Development (UNCED) and the 1994 International Conference on Population and Development. He is currently a senior fellow at Battelle's Pacific Northwest National Laboratory in Washington D.C.

Brack, Duncan
Mr. Brack is a Senior Research Fellow at the Royal Institute of International Affairs (London, U.K.) where he is responsible for work on the interaction of trade liberalization and environmental protection.

Brunner, Walter
Dr. Walter Brunner is a consultant on halon regulations. He also manages the halon register and halon bank for the Swiss government. He is a Co-Chair of the Halons Technical Options Committee and a member of the Technology and Economic Assessment Panel (TEAP).

Buxton, Victor
Mr. Buxton was chief Canadian negotiator for the Montreal Protocol in 1987 and for the 1992 United Nations Conference on Environment and Development (UNCED). He also served as first Chair of the UNEP Technology and Economic Assessment Panel (TEAP) and was a Canadian representative to the Multilateral Fund for the Montreal Protocol.

Carvalho, Suely Machado
Dr. Carvalho is Co-Chair of the Technology and Economic Assessment Panel (TEAP). She has been working for the Montreal Protocol Unit of the United Nations Development

Programme (UNDP) since August 1997, and, until recently, was Director of Technology Development and Transfer for the São Paulo state environmental agency (Brazil).

Cook, Elizabeth
Ms. Cook is a Senior Associate in the Climate, Energy, and Pollution Program at the World Resources Institute (WRI). Prior to joining WRI, Ms. Cook directed Friends of the Earth-U.S.A.'s Ozone Layer Protection Campaign.

Corona, Jorge
Mr. Corona has participated in issues related to the Montreal Protocol since 1990 as a member of the UNEP Technology and Economic Assesment Panel, Co-Chair of the Solvents Technical Options Committee, and member of the Methyl Bromide Technical Options Committee. Mr. Corona is also collaborating with Working Group II of the Intergovernmental Panel on Climate Change.

De Fabo, Edward C.
Dr. De Fabo is a researcher and Professor of dermatology at the George Washington University Medical Center (Washington D.C.). He has worked on the impacts of increased UV-B radiation on human health and the biosphere and is now leading research projects in this subject for the International Arctic Science Committee (IASC) and the International Science Advisory Committee for the Scientific Committee on Problems of the Environment (SCOPE).

Dobó, László
Mr. Dobó has been senior advisor on environmental and other chemical industry issues for the Hungarian Ministries for Industry and Trade, and for the Environment. He has also participated in various negotiations under the Geneva Convention on Long-Range Transboundary Air Pollution. In 1994, Mr. Dobó was selected as a senior advisor to UNEP's Technology and Economic Assessment Panel (TEAP) for the Montreal Protocol.

Fujimoto, Yuichi
Mr. Fujimoto is a senior advisor to UNEP's Technology and Economic Assessment Panel (TEAP) for the Montreal Protocol and member of the Solvents, Coatings, and Adhesives Technical Options Committee. He is also an advisor to the Japan Industrial Conference for Ozone Layer Protection (JICOP).

Kerr, James B.
Dr. Kerr is head of Environment Canada's Stratospheric Ozone Research and Monitoring program. He is the co-inventor of the Brewer ozone spectrophotometer and the UV Index™. He served as member of the International Ozone Commission (1984-1992).

Kerr, Margaret C.
Dr. Kerr is Senior Vice President, Employee and Customer Value for Nortel (Northern Telecom). She chairs the ISO/TC 207 Environmental Management Committee, is a Director of Arthur D. Little Inc., and a member of the Advisory Boards to the World Environment Center and the Royal Society of Canada.

Khosla, Ashok
After having set up the Office of Environmental Planning and Coordination in India's Ministry of Science and Technology, Dr. Khosla became Director of Infoterra, UNEP's global information network. He is now President of Development Alternatives.

Kuijpers, Lambert
Since 1993, Dr. Kuijpers has been an independent consultant doing work on refrigeration and alternative technologies, and technology transfer for, among others, the World Bank, the Multilateral Fund Secretariat, UNEP, and various developing countries. He has co-chaired the Technology and Economic Assessment Panel (TEAP) for the Montreal Protocol since 1992 as well as its Technical Options Committee on Refrigeration, Air Conditioning and Heat Pumps since 1988.

Le Prestre, Philippe G.
Dr. Le Prestre is Professor of Political Science at the Université du Québec à Montréal (UQAM) and Director of Graduate Studies at UQAM's Institute of Environmental Studies.

Mateos, Juan Antonio
Mr. Mateos is Assistant Undersecretary for International Cooperation for the Mexican Ministry of Foreign Affairs. He served as Vice-Chairman and Chairman of the Interim Multilateral Fund of the Montreal Protocol and was head of the Mexican Delegation for the negotiations of the U.N. Convention on Biological Diversity.

McBean, Gordon
Dr. McBean is Assistant Deputy Minister of the Atmospheric Environment Service of Environment Canada, Chairman of the Coordinating Committee for the World Climate Program and a member of the Executive Council of the World Meteorological Organization.

Molina, Mario
Professor Molina currently holds a joint appointment between the Departments of Earth, Atmospheric and Planetary Sciences, and of Chemistry, at the Massachusetts Institute of Technology (MIT). He received, with F. S. Rowland and P. Crutzen, the Nobel Prize in Chemistry in 1995.

Morehouse, E. Thomas
Mr. Morehouse is an Adjunct Research Staff Member at the Institute for Defense Analyses (IDA), a Federally Funded Research and Development Center. He is a member of UNEP's Technology and Economic Assessment Panel (TEAP) of the Montreal Protocol. He co-chaired the Halons Technical Options Committee from 1988 until 1996 and was a military advisor to the U.S. delegation to the Montreal Protocol negotiations in 1987. As Chief of Air Force Pollution Prevention, Mr. Morehouse developed and oversaw implementation of pollution prevention policies worldwide.

O'Toole, Anne
Ms. O'Toole is Director of Environmental Services for the Ontario Region Office of Environment Canada. She led the development and implementation of the UV Index™ program for Canada.

Parson, Edward A.
Dr. Parson is Associate Professor of Public Policy at the John F. Kennedy School of Government at Harvard University and a Senior Research Associate in Harvard's Belfer Center for Science and International Affairs (BCSIA).

Pons, José I.
Mr. Pons is the President of Spray Química C.A., a Venezuelan company. He is a member of the Technology and Economic Assessment Panel (TEAP) and Co-Chair of its Aerosols, Sterilants and Miscellaneous Uses Technical Options Committee.

Rand Sally
Ms. Rand has been at the U.S. Environmental Protection Agency since 1991. She currently serves as the foam sector specialist for the Significant New Alternatives Policy (SNAP) program on the assessment of substitutes for ozone-depleting substances. She is a member of the Technology and Economic Assessment Panel (TEAP) and Co-Chair of its Technical Option Committee for Flexible and Rigid Foams.

Reid, John D.
Dr. Reid is Scientist Emeritus with Canada's Department of the Environment. He was previously Director of Policy and International Affairs in the Department in which role he negotiated the first amendment and adjustments to the Montreal Protocol. He also serves as President of the Canadian Meteorological and Oceanographic Society.

Sand, Peter H.
Dr. Sand is currently a Lecturer in environmental law at the University of Munich. He was responsible for the legal drafting of the 1985 Vienna Convention on the Protection of the Ozone Layer and served as Chief of the Environment Law Unit of UNEP.

Singh, Lalitha
Mrs. Singh is an Adviser in the Department of Chemicals and Petrochemicals of the Indian Ministry of Chemicals and Fertilisers.

Soroos, Marvin S.
Dr. Soroos is Professor and Head of the Department of Political Science and Public Administration at North Carolina State University.

Széll, Patrick
Mr. Széll is Head of the International Environmental Division of the Department of the Environment, Transport and the Regions (U.K.). He has been Chairman of the Protocol's Legal Drafting Group since 1990 and chaired the Working Group which drew up the Protocol's non-compliance procedure between 1989 and 1992.

Taylor, Gary
Mr. Taylor is a member of the UNEP Technology and Economic Assessment Panel (TEAP) for the Montreal Protocol and Co-Chair of its Halons Technical Options Committee.

Tevini, Manfred
Professor Tevini teaches in the Department of Botany at the University of Karlsruhe (Germany). He was a member of the Coordinating Committee for the Protection of the Ozone Layer and co chaired the UNEP Environmental Effects Panel for the Montreal Protocol.

Tolba, Mostafa K.
Dr. Tolba is currently President of the International Center for Environment and Development in Cairo (Egypt) and a Professor in the Faculty of Science at Cairo University. He served as Executive Director of the United Nations Environment Programme (UNEP) from 1976 to 1992.

Tope, Helen
Dr. Tope has worked since 1991 for the Environmental Protection Authority of Victoria (Australia) on issues including ozone protection, air emissions inventories, and emergency response. She is a member of the Technology and Economic Assessment Panel (TEAP) and co-chairs its Aerosols Technical Options Committee.

van der Leun, Jan C.
Dr. van der Leun is Emeritus Professor of Dermatology at Utrecht University Hospital (Netherlands) and co-chairs the UNEP Environmental Effects Panel for the Montreal Protocol since 1988. He was chairman of the Effects section of the UNEP Coordinating Committee on the Ozone Layer between 1982 and 1988.

Van Slooten, Robert
Dr. Van Slooten has been involved in various task forces and committees under the Montreal Protocol and is currently the Co-Chair of the Economic Options Committee (EOC) of UNEP's Technology and Economic Assessment Panel (TEAP) for the Montreal Protocol as an independent consultant on contract to the U.K. Department of the Environment.

Woodcock, Ashley
Dr. Ashley Woodcock is a Consultant Respiratory Physician at the North West Lung Center, Wythenshaw Hospital, Manchester (U.K.). He is a full-time practising physician and a Senior Lecturer at the University of Manchester.

Worrest, Robert
Dr. Worrest is a member of the UNEP Environmental Effects Assessment Panel for the Montreal Protocol and Director of Washington Operations for the Consortium for International Earth Science Information Network (CIESIN).

Acknowledgements

This book, the product of the Tenth Anniversary Colloquium of the Montreal Protocol that took place in September 1997, would not have been possible without the financial support from Environment Canada and from University of Quebec at Montreal's Publication Committee, as well as the help of many individuals who gave their time to assemble this distinguished array of expertise and experience. In particular, we wish to thank the members of the Tenth Anniversary Colloquium's International Advisory Committee: Pieter Aucamp, Suely Machado Carvalho, John Hollins, Winfried Lang, Nelson Sabogal, Jan C. van der Leun, and the other members of the National Organizing Committee: Angus Fergusson, Sonja Henneman, Claude Lefrançois, Yarrow McConnell, Sara Melamed, Robert Saunders and Hague Vaughan. Darcy Longpré and Noémie Pelzer transcribed the presentations. François Nicolas Pelletier, assisted by Geneviève Reed, oversaw the production of the manuscript with great dispatch and efficiency, using the facilities generously provided by the Institute of Environmental Sciences of the University of Quebec at Montreal. The views in this book are those of the individual authors and do not necessarily reflect those of their organizations, the sponsors of the Colloquium, or the editors of this publication.

List of Acronyms

AES	Atmospheric Environment Service (Environment Canada)
APC	Antigen-Presenting Cells
ASEAN	Association of South-East Asian Nations
CANZ	Canada, Australia, New Zealand
CART	Classification and Regression Tree
CBD	Convention on Biological Diversity
CCOL	Coordinating Committee on the Ozone Layer (UNEP)
CEIT	Countries with Economies in Transition
CFC	Chlorofluorocarbon
CHS	Contact hypersensitivity
CIS	Commonwealth of Independent States
CITES	Convention on International Trade in Endangered Species of Wild Fauna and Flora
COP	Conference of the Parties (Vienna Convention)
DoD	Department of Defense (U.S.)
DoE	Department of Energy (U.S.)
DU	Dobson Units
ECE	Economic Commission for Europe (U. N.)
EFTA	European Free Trade Association
EIAJ	Electronic Industries Association of Japan
EMS	Environmental Management System
ENGO	Environmental Non-Governmental Organization
EPA	Environmental Protection Agency (U.S.)
E.U.	European Union
FCCC	Framework Convention on Climate Change
FEPC	Federation of Electric Power Companies (Japan)
FPI	Foodservice and Packaging Institute (U.S.)
GATT	General Agreement on Tariffs and Trade
GAW	Global Atmospheric Watch
GDP	Gross Domestic Product
GEF	Global Environment Facility
GNP	Gross National Product
HARC	Halon Alternatives Research Corporation
HCFC	Hydrochlorofluorocarbon
HFC	Hydrofluorocarbon
IASC	International Arctic Science Committee
ICCP	International Climate Change Partnership
ICEL	International Cooperative for Environmental Leadership
ICOLP	Industry Cooperative for Ozone Layer Protection
IGY	International Geophysical Year
IPCC	Intergovernmental Panel on Climate Change
JAMA	Japan Automobile Manufacturers Association
JEMA	Japan Electrical Manufacturers Association

JFMA	Japan Facility Management promotion Association
JICOLP	Japan Industrial Conference on Ozone Layer Protection
JRAIA	Japan Refrigeration and Air Conditioning Industry Association
LRTAP	Long-Range Transboundary Air Pollution
MCP	Multilateral Consultative Process
MEA	Multilateral Environmental Agreement
MITI	Ministry of International Trade and Industry (Japan)
MLF	Multilateral Fund
MOP	Meeting of the Parties (Montreal Protocol)
NASA	National Aeronautics and Space Administration (U.S.)
NGO	Non-Governmental Organization
ODA	Official Development Assistance
ODP	Ozone-Depleting Potential
ODS	Ozone-Depleting Substance
OECD	Organization for Economic Cooperation and Development
OEWG	Open-Ended Working Group
OTP	Ozone Trends Panel
PFC	Perfluorocarbon
PVC	Polyvinyl Chloride plastics
RIA	Regulatory Impact Analysis
SBI	Subsidiary Body on Implementation
SCOPE	Scientific Committee on Problems of the Environment
TEAP	Technology and Economic Assessment Panel
TOC	Technical Options Committee
UCA	Urocanic Acid
U.K.	United Kingdom
U.N.	United Nations
UNCTAD	United Nations Conference on Trade and Development
UNDP	United Nations Development Programme
UNEP	United Nations Environment Programme
UNIDO	United Nations Industrial Development Organization
U.S.	United States
UVIRC	Ultraviolet International Research Center
VOC	Volatile Organic Compounds
WMO	World Meteorological Organization
WOUDC	World Ozone and Ultraviolet Radiation Data Centre
WTO	World Trade Organization
WUDC	World Ultraviolet Radiation Data Centre
ZA	Solar zenith angle

PROTECTING THE OZONE LAYER:

Lessons, Models, and Prospects

THE MONTREAL REGIME: A NEW MODEL FOR INTERNATIONAL COOPERATION ON GLOBAL ENVIRONMENTAL ISSUES?

Philippe G. Le Prestre, John D. Reid and E. Thomas Morehouse

The Montreal Protocol to the 1985 Vienna Convention for the Protection of the Ozone Layer, signed on September 16, 1987, is one of a small number of international environmental agreements that have had rapid and concrete impacts on the actions of nations and other groups responsible for the quality of the environment. The Protocol was signed at the dawn of a period of growing interest in global environmental issues. None of the subsequent international agreements has been as successful as the Montreal Protocol in creating the incentives and mechanisms for change, in part due to different political, scientific and technological contexts, to the nature of the environmental issue itself, and to their relative newness. Many of them, however, contain features already present in the Montreal Protocol, and changes in their operations as well as specific protocols to be negotiated within these regimes could well benefit from this experience.

To mark the tenth anniversary of the signing of the Montreal Protocol, a series of events were held in Montreal in September 1997, in conjunction with its Ninth Meeting of the Parties. Among them was the Tenth Anniversary Colloquium, a one-day event involving over fifty speakers and attracting more than three hundred participants from around the world. The Colloquium reviewed the roles played by the natural and social sciences and by technology in the development and implementation of the Protocol, presented the state of scientific knowledge, and discussed lessons for strengthening the implementation of the Montreal Protocol and for facilitating co-operation on other global environmental issues, such as climate change. This book is the product of that colloquium. It offers a rare combination of the experience and perspectives of practitioners and observers from three different fields—the natural sciences, the social sciences, and technology—that have contributed significant knowledge and innovative mechanisms toward the protection of the ozone layer.

While the 1985 Vienna Convention only urged States to adopt measures to reduce their consumption of harmful chemicals, Parties to the Montreal Protocol agreed to reduce consumption of key chlorofluorocarbons (CFCs) to fifty percent of 1986 levels by 1998. Through the efforts of industry, government and public interest groups, and motivated by improvements in scientific understanding, technical capability, and a willingness to overcome social and economic barriers, reductions in use and phase-outs have progressed further and faster than expected while the list of controlled chemicals has expanded. At meetings in London in 1990 and Copenhagen in 1992, Parties accelerated the original reduction schedules and added new substances to the list. Three years later, at the Seventh Meeting of the Parties in Vienna, Parties agreed to phase out methyl bromide.

Between 1986 and 1994, the use of ozone-depleting substances decreased by ninety percent in Organization for Economic Cooperation and Development (OECD) countries and, as Mario Molina points out, there is evidence that atmospheric concentrations of CFCs have begun to drop although ozone is still being depleted. On the other hand, the use of CFCs and halons in some major developing countries (such as India, the Philippines, or China) doubled during the same period. Developing countries accounted for roughly sixty percent of the total use of CFCs and halons in 1994. And while the use of chlorine-containing ozone-depleting substances (ODS) is decreasing, that of bromine-based ODS continues to increase. Finally, as of April 1998, 165 countries had ratified the Montreal Protocol, 120 the London Amendment and seventy-eight the Copenhagen Amendment.

These are the substantive accomplishments of the Protocol. Underpinning them are accomplishments in natural science, technology, and policy. In this introduction, we draw out some of the lessons of the past ten years that have made the larger accomplishments possible and may provide guidance for other international negotiations on global environmental problems. These lessons fall into three different categories: encouraging the development and acceptance of the Protocol, facilitating compliance, and promoting the adaptation of the regime over time.

Development and acceptance

For many participants to the Tenth Anniversary Colloquium, the negotiation and endorsement process have been facilitated by a series of conditions that encouraged governments to take rapid and meaningful action. Of particular significance was the role of science and of scientists—several of whom are among the contributors to this volume—in identifying the problems and assessing the impacts of various proposals on the ozone layer.

Science played a major role in making the Protocol both possible and effective. It provided the information that stimulated concern. Developed over generations, it made rapid action by governments possible, thus, once again, illustrating Isaac

Newton's famous remark: "If I have seen farther it is by standing on the shoulders of giants." Without the discovery of the ozone layer in the nineteenth century, largely out of academic interest, the world would have been faced with a much more serious problem when the impacts of increased ultraviolet radiation became obvious. Without systematic measurements that started during the International Geophysical Year (1957-58), we would have no basis for knowing that ozone layer depletion began only a few years later, most dramatically in the Antarctic. Without the early measurements of CFC concentrations in the atmosphere by James Lovelock, there would have been no stimulus to find the sink for these chemicals in the ozone layer. Many scientists at the Colloquium, particularly those dealing with impacts of ultraviolet radiation, expressed concern that the current generation of science will not be leaving a similar legacy of understanding for future generations because of a lack of vision and of the domination of short-term thinking in allocating funding to environmental science.

Science alone is not enough, it should also be successfully communicated. And so it was. Mostafa Tolba and Richard Benedick, among others, point out that the mobilisation of the public by the scientific community, the media, and concerned groups helped put the issue on the agenda and pressure governments. Graphic satellite images of the Antarctic ozone hole brought home its reality. U.S.-led experiments that showed an anti-correlation between ozone and ozone-depleting chemicals inside and outside the Antarctic vortex, were described as a "smoking gun". The World Meteorological Organization (WMO) provided timely bulletins on the development of each year's ozone hole. As described by Anne O'Toole, a daily UV Index—developed in Canada and subsequently adopted as an international standard—sensitises people to the day-to-day variation in ultraviolet radiation hazard. These communications helped build public understanding and support for international action.

Largely on the basis of the science, a core group of countries—among them Canada and Scandinavian countries, and later the United States—and strong personalities—notably Mostafa Tolba, then Executive Director of the United Nations Environment Programme (UNEP)—pushed for action. UNEP, in particular, adopted a remarkably pro-active role. As Mostafa Tolba recalls, this dynamism was facilitated by the existence of clear scientific models. Later, corroborating findings sustained the initial commitments and the adoption of procedures designed to integrate this information into the negotiations, and ensured that it informed the on-going diplomatic process. Moreover, according to Richard Benedick, clear signals and a stable regulatory environment encouraged industry to invest in new technologies. In the final analysis, public policy pronouncements and incremental funding cannot solve a problem which is fundamentally technological in nature. According to Victor Buxton, the architects of the Montreal Protocol recognized that over time scientific understanding of the process of ozone depletion and the technologies available to halt the depletion would fundamentally influence the policy options available to the Parties.

The structure of the regime also facilitated the participation of a large number of countries. The trade provisions, as Duncan Brack concludes, were vital in building the wide international coverage it has achieved and in preventing industrial migration to non-parties in order to escape the controls on ODS. Yet, as Peter Sand reminds us, China and India would never have joined the Montreal regime without the adoption of a new and innovative financial mechanism—the Multilateral Fund—and the acceptance of the principle of differentiated responsibilities which allows some countries to pay for others to comply later since, as Ashok Khosla bluntly puts it, poorer countries had no choice but to join.

Compliance

Of course, a problem is not solved merely by signing agreements or banning products. As Patrick Széll notes, the Montreal Protocol compliance model, with its potentially intrusive scrutiny, is unique, at least in environmental regimes. The compliance mechanisms included in the Montreal Protocol are broadly worded and have not yet been formally invoked. The details by which compliance will be enforced have not yet been formulated, mainly because they have not been needed. While there are issues with Russia, developing countries, black marketing, and methyl bromide, the Parties are making progress toward resolving them without the need to resort to punitive measures. Efforts have focused on assisting Parties to achieve compliance. For example, through the application of the essential use process and the development of country programs, the countries with economies in transition (CEIT) are being brought into compliance.

Non-compliance is indeed frequently the result of a lack of capacity, rather than of will, to comply, and thus a global capacity-building process is required as part of any international environmental agreement. Peter Sand notes that the Montreal regime has evolved both carrots and sticks to encourage compliance, the stick being the formal compliance procedure and the carrot being the incremental cost subsidies provided through the Multilateral Fund. In addition, as Edward Parson, James Armstrong and Elizabeth Cook all agree, the cost and difficulty of meeting phase-out schedules of ODS under the Protocol have often turned out to be substantially smaller than were initially predicted. Indeed, it is important to put the current situation with respect to the progress made by the Protocol in context with the arguments that were being made against the Protocol at the time of its signature. None of these claims, such as the costs of conversion, public safety hazards, national security impacts, or having to rely on inferior products and technologies have turned out true. Yet, these same arguments have again surfaced in the discussions about climate protection.

The Montreal Protocol is widely hailed for the way it has negotiated compliance; but compliance is also linked to the design of the regime and to the mechanism for dispute settlement. First, the agreement identified specific time-bound targets, then it assessed the required funding needs and sources, and then it set up a standing Implementation Committee which is a forum where a fair plan for compliance can be worked out (Victor 1996, Széll 1996). Patrick Széll identifies its sensitivity to the particular concerns, needs and fears of the Parties as one of the most notable operating features of the Implementation Committee. Moreover, it was allotted time for the definition of its tasks; it defined its role outside the basic legal text, initially in a modest and flexible matter; it was composed of a limited number of state representatives serving stable terms, was subservient to the Meeting of the Parties, and, above all, was conceived not as a quasi-judicial dispute settlement mechanism but rather as an assisting body that contributes to international consensus. In this context, non-compliance procedures have been most effective when they operated on a case-by-case basis and maintained clear links to the financial mechanism (Victor 1996). Yet, flexibility has its limits. Peter Sand, for example, expresses concern with the Implementation Committee's recent practice of finding certain countries in compliance just so they can remain eligible to receive incremental cost subsidies through the Multilateral Fund. Such "consensual redefinition of treaty standards", according to him, could weaken its effectiveness. Did success, then, lie more in finding accommodations to non-compliance than in encouraging compliance? Or, as some would argue, did short-term accommodations lay the ground for long-term compliance?

Consensus was also a hallmark of the operation of most of the Technical Options Committees (TOCs) even when initial apprehensions were high. While halon producers were strongly represented in the Halons TOC and strongly opposed to controls, according to Walter Brunner, the overall service orientation rather than product orientation of the industry ultimately resulted in a consensus on halon phase-out. Service providers were able to continue providing comparable fire protection services without halons and found business opportunities in alternatives. It is notable that the military was well represented in the Halon TOC and provided strong leadership in the development and implementation of halon alternatives. They also pioneered the implementation of "banking" of ODS for important uses, which was first proposed by Gary Taylor. Ashley Woodcock also noted that the essential use process provided relief for small but important uses until alternatives could be developed, such as for CFC-based metered dose inhalers used by asthma patients.

Implementation of the technologies identified by TEAP and its TOCs was also facilitated by individuals from these same bodies through international co-operation projects. Margaret Kerr and Yuichi Fujimoto present examples of multinational companies, in many instances competitors, co-operating to develop and share alternatives technologies world-wide. Yuichi Fujimoto explains how Japanese and Thai companies co-operated with the Thai government to achieve the first complete phase-out in a developing country in an industry sub-sector, in this case domestic refrigeration. In this spirit, Stephen Andersen describes how the TEAP process produces individual "champions" who advanced the cause of ozone protection within companies and governments, and, through individual initiative, launched significant projects which demonstrated that phasing out ODS was not as difficult or costly as originally believed.

Issues of compliance require innovative solutions. But, as Edward Parson suggests, "an adaptive regime must be able to tolerate less than universal compliance without unravelling. A regime that remains functional only with perfect compliance will either break apart or come progressively to be re-defined so that compliance is meaningless. The requirement is to maintain commitment sufficient to force real effort, without being so rigid that a single instance of failure brings the regime down." But where does flexibility end and unravelling begin? Mostafa Tolba stresses that flexible compliance, to be truly effective and pregnant of future progress, must be accompanied by well-developed and well-accepted procedures for the verification of the contracted obligations. Hence the importance of national data reporting and of a system of performance reviews of individual parties. Not only can it prod laggards, it can help assess and compare performances as well as incite domestic groups to pressure their respective governments. Still, by June 1997, only 105 of the 153 Parties required to do so had reported data for 1995. Only twenty-one Parties had so far reported data for 1996. Russia, Ukraine and Japan experience severe compliance difficulties. László Dobó and Lambert Kuijpers show how the TEAP has approached this difficult question. In particular, they credit TEAP with assembling the first set of consumption data for CEIT and playing a key role in the Parties' decisions to provide specific assistance to CEIT

Adaptation

As Mostafa Tolba emphasises, environmental agreements will be successful if they are flexible and contain the capacity to adapt to change in their scientific and political context. In this case, the development of a mechanism to integrate new scientific findings into the evolution of the regime proved crucial. It is on the basis of this continuous assessment process that Edward Parson can call the ozone regime "the world's first adaptive global environmental regime".

The political, scientific and technological contexts of agreements has evolved over time. One of the most remarkable successes of the Protocol, one that has subsequently been adopted for the global climate change and biodiversity conventions, is science assessment. In his insightful remarks, Daniel Albritton, points to some of the factors that made assessments so successful in driving change in the Protocol. One major contribution was to keep debate about the science within the expert domain, subject to thorough international peer review, to document the details but also to report findings in a form digestible to the non-specialist—the summary for policy makers. This keeps the science at arms length from negotiation. It frees the Protocol negotiators, many of whom are unqualified to deal with, and even intimidated by, the science, to wrestle with the policy implications. Nevertheless, some argue that the assessment process stifles dissenting scientific viewpoints. The credibility of the assessment process will only be maintained to the extent that it continues to live by strong peer review and that institutions recognise that a scientist's contribution to assessments is worth rewarding.

Technological evolution was also able to inform the evolution of the regime. One of the key factors in the success of the Montreal Protocol was the reliance on a technology assessment mechanism and the structure of the TEAP organization. Technical experts from stakeholder organizations, nominated by national governments and independent of external instruction, conduct objective assessments of the technical and economic feasibility of new technologies. The results of the TEAP form the basis for specific projects that can assess the viability of proposed alternatives, and their results produce wider implementation.

Article 6 of the Montreal Protocol required periodic assessments of the control measures on the basis of available scientific, environmental, technical and economic information. This drove the evolution of a series of assessment panels and options committees. Suely Carvalho and Robert Van Slooten describe the structure of the Technology and Economic Assessment Panel (TEAP) and its seven Technical Options Committees (TOCs). Suely Carvalho attributes the success of these bodies to an era of increasing corporate environmental responsibility and to the volunteer nature of the TEAP and its TOCs which attract experts with particular environmental interests. As a result, these experts tend to be focused on finding technologies that work rather than finding reasons why technologies may not work. According to Van Slooten and Radhey Agarwal, some of the key operating principles of these bodies include the policy relevance of their technical reports, the independence of its members from operating "under instructions", the global composition of the body of experts, and its consensus-based decision making process. As Helen Tope points out, the independence of the bodies is important because it separates technological considerations from commercial or political interests. Sally Rand and Lalitha Singh provide insights regarding the importance of TEAP in fostering international technology co-operation among industry and government, and creating a community of "technology ambassadors" in the six hundred volunteer experts from forty-five countries who have participated in the TEAP process.

The operation of the TEAP and its TOCs is not without challenges, however. Jonathan Banks, Co-Chair of the Methyl Bromide Technical Options Committee reports polarisation of opinions within the committee, driven largely by commercial interests. Strong participation by chemical producers which have no interests in developing alternatives, coupled with a highly risk averse community of users is serving to destabilise the process. Indeed, the Methyl Bromide TOC has been the most problematic of the TOCs in terms of achieving consensus. Unfortunately, many of the same companies which produce methyl bromide also produced halons, another family of brominated compound, which were the first ODS to be phased out in developed countries under the Protocol.

Adaptation involves integrating new knowledge not only about the problem at stake but also about the links among environmental issues (e.g. ozone depletion, climate change and acid rain) and about how human societies mediate the consequences of environmental change. Thus, one should avoid the pitfalls of determinism and move away from a conception of the problem that sees a direct and automatic link between ecosystems and societies. Further, little thought has been given to potential trade-offs between regimes. For example, as Mostafa Tolba and Dan Albritton point out, ozone negotiations brought to the surface the close links between protecting the ozone layer and controlling climate change. What could be a solution for one convention could be a problem for another. Controversies regarding the use of trade sanctions and the relationship between environmental and trade regimes have yet to be settled. In the latter case, as Peter Sand warns us, governments and the World Trade Organization (WTO) may be more worried about secondary, "regime-born", trade barriers—developed as in the case of the 1973 *Convention on International Trade in Endangered Species of Wild*

Fauna and Flora (CITES) only *after* the treaty had been negotiated and adopted—than about the original trade-related provisions found in the text of multilateral environmental agreements.

What kind of model?

A few observers and practitioners have already reflected on the lessons that the ozone regime process could hold for environmental co-operation in general, and climate change in particular (e.g., Lang 1991, 1996). One of the reasons for doing so is to probe which aspects of the Protocol could offer guidance and inspiration to other international environmental agreements. At the time of the tenth anniversary of the Protocol, climate change was most prominent in the minds of the delegates and participants to the Colloquium. The lessons drawn from the experience of the Montreal regime are many, but not all observers or participants will draw the same implications from them.

Every scientist, diplomat, lawyer or technical expert who has been closely associated with the development and evolution of the Montreal Protocol tends to emphasise the particular contribution of his field. Moreover, those who lived through the ups and downs of difficult negotiations are apt to stress what has been achieved. As Victor Buxton testifies, given the lack of scientific or economic consensus on the key issues surrounding ozone depletion at the time of the Montreal Protocol negotiation, and given the differences within and among the E.U., the U.S. and the developing world, agreement was a hard-won battle. Some observers have not hesitated to claim that the devices used for protecting the ozone layer could also be used for other treaties, most likely climate change. Mostafa Tolba, a prime mover behind the agreement, believes that the Climate Change, Biodiversity, Desertification, and Basel Conventions could all benefit greatly from what he considers a model of international co-operation. Richard Benedick underscores innovative elements within the negotiation process, the structure of the Montreal regime, the faith in market mechanisms, sensitivity to equity issues, and the development of a world-wide network of NGOs. Although Patrick Széll is more cautious, noting that, in the case of climate change, the need appears to be for an advisory rather than a supervisory multilateral consultative process, Peter Sand points out that, like it or not, some of its features crop up in other agreements. Are we witnessing, he asks, a "process of inter-organisational social learning"? If so, what are the dimensions and limits of that learning?

Science was crucial in providing the basic model, in mobilising certain governments, and in demonstrating the impacts of ODS on the ozone layer and human health. Successive new scientific findings about the deterioration of the ozone layer, given the stamp of approval by the Protocol's own international expert assessment process, continued to motivate strengthening the agreement. Indeed, information on social and ecosystemic impacts in areas other than the Antarctic were lacking at the time of negotiation and amendment of the Protocol.

Yet, Parties were not always so sure as to the applicability of any lessons, or willing to entertain that idea. The U.S., for example, has tended to believe that the ozone and climate change are such different issues that one cannot learn from the ozone process lessons that would be relevant to carbon dioxide. Ashok Khosla, while acknowledging the success of the Montreal regime, vividly points to these differences. The Montreal Protocol deals with a relatively well-defined problem which requires mostly technical solutions rather than major lifestyle changes. This is not necessarily the case with climate change or biodiversity. His image of a negotiating game where the rules are written and changed at will by certain players, and which other players are not allowed to quit, illustrates powerfully the perception that leaders of some developing countries may have about the global environmental negotiating game. These perceptions are worth bearing in mind as the international community turns to other problems. This pessimism about the transferability of lessons learned from ozone protection is shared by Marvin Soroos who emphasises the importance of domestic politics and that adaptation—rather than prevention—may be a more appealing strategy to governments in the case of climate change.

Indeed, the applicability of lessons is one question; another is the willingness of governments to go down the same path. For example, MLF-style arrangements have been explicitly rejected in the case of biodiversity or climate change. Even the precautionary principle has not been as universally accepted as one would expect. Moreover, crucial tests regarding the willingness and capacity of developing countries to respect their obligations under the treaty have yet to be passed.

It is often difficult to disentangle the relative contributions of the factors that have been identified with the success of a regime. In part, this is because an international agreement is a balanced package where each Party may perceive the importance of various elements differently. How, then, can we isolate the importance of, say, trade restrictions as opposed to the faster than expected availability of cheaper substitutes? Of course, if the issue is banning certain substances, then trade sanctions may be important. In principle, Duncan Brack believes, trade measures may have an important role to play in other multilateral environmental agreements (MEAs). Their precise form will of course vary with the MEA in question, and in some cases they are likely to be more politically credible and technically feasible than in others. For example, they may be more easily applicable to an agreement controlling persistent organic pollutants than to the Climate Change Convention. And trade sanctions should always be accompanied by effective finance and technology transfer mechanisms if the MEA is to be regarded as fair.

Even though the tightening of restrictions and widening of the range of products under phase-out or production limits has been a remarkable feature of the evolution of the Montreal Protocol, the challenges ahead are many. Scientific studies on the impacts of ozone depletion are still embryonic. Smuggling needs to be checked—initial steps to that end were adopted at Montreal in 1997—and compliance improved in certain areas. Difficult trade-offs lie ahead, as shown by the case of methyl bromide. But the biggest challenges will be, first to maintain a unique process that allows the Parties to adjust to new information and, second, to ensure that the lessons that arise from that process are accepted and allowed to inform other attempts to deal with global issues. One may not like the lessons that can be drawn from it or even find them all relevant, but one cannot ignore them.

References

Lang, Winfried. 1991. "Is the Ozone Depletion Regime a Model for an Emerging Regime on Global Warming?" *UCLA Journal of Environmental Law and Policy* 9: 161-74.
Lang, Winfried, ed. 1996. *The Ozone Treaties and their Influence on the Building of Environmental Regimes.* Austrian Foreign Policy Documentation. Vienna: Austrian Ministry of Foreign Affairs.
Széll, P.J. 1996. "Implementation Control: Non-Compliance Procedure and Dispute Settlement in the Ozone Regime." In *The Ozone Treaties and Their Influence on the Building of International Environmental Regimes.* Austrian Foreign Policy Documentation. Austrian Ministry of Foreign Affairs.
Victor, David. 1996. "The Montreal Protocol's Non-compliance Procedure: Lessons for Making Other International Environmental Regimes More Effective." In Winfried Lang, ed., 1996: 58-81.

PART 1

THE SCIENTIST AND THE POLITICIAN

STRATOSPHERIC OZONE CHEMISTRY

Mario Molina

It is a pleasure to be with you this morning to celebrate the tenth anniversary of the Montreal Protocol. I know that most of you are familiar with the science related to the stratospheric ozone depletion issue, so I will just present a short review of some of the scientific community's accomplishments, as well as my perspective on the future. The accomplishments were quite extraordinary, developing a very strong case that facilitated the Montreal Protocol negotiations. Of course, not only were the scientific aspects successful, but so were the negotiations and the diplomacy.

Let me go back in time to put briefly this problem in perspective. This was an important issue as it clearly showed us for the first time how human activities can affect the entire globe. We have known for centuries that humans can pollute their immediate environment, but this was the first example demonstrating that human-generated pollution can reach global proportions. The ozone problem also illustrated how fragile the atmosphere is. It is really a very thin layer: most of the mass of the atmosphere is confined to the first 20 km; compare that to the distance between the two poles, which is 20,000 km.

Development of the Issue

When did this issue start? We can examine an example of the impact technological advances have had on the atmosphere by going back about a century. Around 1910, home refrigerators began to be very popular. They were quite primitive; they were merely ice boxes. Technology advanced, and refrigerators, more or less as we know them today, were developed. However, these early refrigerators had some severe problems: the original cooling fluids, sulfur dioxide or ammonia, were rather toxic. Several serious accidents occurred, with families living in small rooms that experienced refrigerator leaks. This matter caught the attention of several people who decided to solve the problem, some of whom were actually rather prominent: Albert Einstein, working with Leo Szilard, developed a number of alternate refrigeration cycles which, however, never became commercially successful owing to their complexity, and perhaps also to the Great Depression. What actually succeeded were refrigerators similar to those of the 1930s, with the exception of the toxic chemicals. These were now replaced by new "miracle" chemicals developed by Thomas Midgley, a mechanical engineer, turned chemist. Midgley systematically examined the periodic table, deciding which elements to combine in order to produce very stable compounds, and created the chlorofluorocarbons, better known as CFCs. Midgley also developed other important chemicals, such as the anti-knock agents used in automobile engines, consisting of tetraethyl lead. In his time, Midgley never became aware of the environmental problems these chemicals would eventually generate.

The CFCs have two very important sets of properties. One is that they are fluids readily transformable from liquids to vapors and vice versa, at close to normal atmospheric conditions; this is a very important property that makes them suitable as refrigerants. The second important property is their extreme stability and low toxicity. It is for these reasons that the CFCs became so popular in other uses as well, leading to a rapid increase in their production during the 1960s and 1970s; for example, they were used extensively as propellants for aerosol spray cans.

I would like to summarize the subsequent developments very briefly. In 1974, my colleague Sherwood Rowland and I suggested the CFC-ozone depletion theory. What happens is that the CFCs released to the environment are very stable, and do not decompose in the lower atmosphere—the troposphere. Normally the troposphere's cleansing mechanisms eliminate reactive or water-soluble compounds quite quickly, but the CFCs survive their trip up to the stratosphere—which has no such self cleansing mechanisms—and where the ozone layer, which protects life from harmful ultraviolet radiation coming from the sun, resides. The CFCs finally degrade in the stratosphere, which, however, is also very resistant to vertical movements. Hence, the very reactive components produced by the decomposition of the CFCs stay in the stratosphere for an extended period of time, participating in catalytic cycles of ozone destruction.

The concept of catalytic cycles, developed by Paul Crutzen and Harold Johnston, explains how it is that very small amounts of certain species, at concentrations in the parts per billion level, can control the ozone that exists in parts per million concentrations. Chlorine is a very efficient catalyst for this ozone destruction process. Many assumptions of this theory were tested by measurements made in the laboratory and in the atmosphere, but it was not until 1985 that the first indications of atmospheric effects on ozone itself were reported by Joseph Farman and his colleagues. These measurements showed a rapid decline in the ozone over Antarctica during the springtime. Nobody in the scientific community had predicted that ozone depletion would occur specifically over the South Pole. When the data were first presented, there were many conflicting explanations. Some even thought that lower-level, ozone-poor air was displacing the normally ozone-rich layer of the stratosphere, and that no chemical destruction was taking place.

However, the scientific community was able to collaborate and conduct experiments designed to reach a definitive conclusion. To highlight one of these experiments: a heavily instrumented ER-2 aircraft flew right through the region where the ozone depletion was the heaviest, in order to verify and test the various theories. The results were very clear: ozone was being depleted by chemicals related to the CFCs, clearly showing that the depletion was not a natural phenomenon.

We have seen many times how human society can affect the environment. What is unique about Antarctic ozone depletion is not only the large magnitude of the effect, but also that it takes places as far from the source of pollution as is possible; the source being predominantly in the northern hemisphere, and the most important impacts occurring near the South Pole. Indeed, the "ozone hole" has become a symbol of what we can do to the global environment. The effect is so large, the signal is so strong, that it is not difficult to monitor. Measurements of ozone concentrations as a function of altitude made over the South Pole show how ozone decreases with time during the Austral spring: in a matter of several weeks ozone virtually disappears over a 5 km layer of the stratosphere—over ninety-nine percent of the ozone is depleted in that layer.

The reason that Antarctic ozone depletion had not been predicted by the scientific community prior to 1985 is that we had not considered the possibility of cloud formation caused by the extreme cold temperatures at the Poles. It turns out that the cloud particles speed up the production of chlorine catalysts which cause ozone depletion, a process which we understand very well now. In fact, we can now measure the free radical concentrations as well as the ozone levels even from satellites, observing spectacular results with surprisingly large concentrations of the chlorine catalysts responsible for the ozone loss.

Current Issues

Let us now turn to some current issues. You are all probably familiar with the scientific assessments that are a part of the Montreal Protocol process. The last assessment, that of 1994, included a section designed to answer some common questions that many skeptics have brought up over the years and continue to do so. I will address a few of these questions and some of the subsequent developments.

One common question is that with so many natural sources of chlorine (oceans, volcanoes, etc.), it seems surprising that CFCs would be the principal source of chlorine in the stratosphere. We actually have many measurements that show this to be the case, unambiguously. Satellite measurements monitoring concentrations of hydrogen fluoride, hydrogen chloride, and the total organic halogen budget leave no doubt as to the source of the bulk of the chlorine in the stratosphere: it comes from the CFCs, although a natural source, methyl chloride, also contributes, as has been known for many years. We also know that ozone is not only depleted over the poles, but also over the entire globe, particularly at higher latitudes; the decreasing ozone trend is quite clear now even at mid-latitudes.

There have been problems with model predictions. Typically, ozone depletion has been larger that we could account for with our early calculations. We have come a long way towards more realistic models. A number of scientists, like Susan

Solomon, have made very prominent advances in this area. We think that we understand much better the source of the discrepancy. Some of it is related to heterogeneous chemistry and to bromine chemistry; an example is a reaction of bromine nitrate on the sulfate aerosol particles that exist in the stratosphere, a process which speeds up the production of catalysts that affect ozone. Another reason has to do with the non-linear behavior of the chemical system: an example is one of the reactions in the catalytic polar ozone destruction cycle, whose rate depends on the square of the chlorine monoxide concentration. Hence, inhomogeneities—the fact that the atmosphere is not perfectly well mixed—also account for a substantial part of the discrepancies in these calculations. Furthermore, the use of average zonal temperatures within the models distorts results because heterogeneous reactions are particularly efficient at very low temperatures, and so we have to account for the excursions to low temperatures. When these factors are taken into account we can explain most of the discrepancy between earlier models and observations.

Another issue that was raised at one time is the trend in ultra-violet radiation at the earth's surface, which is of course an important concern. While there is a clear trend of ozone loss, the evidence for trends in ultraviolet radiation had been debatable. The problem is that no baseline ultra-violet measurements existed, and so it is difficult to make a case that there has been a change. Of course, over Antarctica, this task is simplified because the change is very large. Recently we have made considerable progress gathering data on ground-level UV radiation that show the expected correlation with ozone depletion. An example is the article by Elizabeth Weatherland and colleagues, which takes another look at Scotto et al. very controversial paper that appeared in *Science* in 1988, which concluded that UV levels were actually decreasing. Weatherland's analysis exposes a number of flaws in that earlier *Science* article; overall, the new analysis shows good correlation of UV increase and loss of stratospheric ozone when other variables like dust, pollution and clouds are properly taken into account.

The Future

Finally, I would like to talk about the future of the CFC-ozone issue. First, regarding chlorine concentrations, measurements already show that global CFC concentrations are leveling off. Concentrations of CFC-12, a long-lived CFC, are no longer increasing nearly as fast as they were a few years ago. For CFC-11, which is shorter-lived, concentrations have already leveled-off. For the even shorter-lived gas methyl chloroform, the concentrations are decreasing rapidly, in response to industry's anticipation of the problem. Putting all this information together, it is evident that chlorine has peaked and is beginning to decrease slightly. This is a sign that the Montreal Protocol is working; there are, of course, problems, but the overall trend is encouraging. We expect concentrations of chlorine to decrease further in the future after the full implementation of the Montreal Protocol, and thus, eventually, the ozone hole should disappear.

The bromine situation is, however, a bit more worrisome. Atom for atom, bromine is about 50 times more efficient in depleting ozone than chlorine. Our understanding of atmospheric bromine chemistry is advancing, but is by no means perfect yet. The concentrations of halons, the main bromine containing ozone-depleting compounds, do not appear to be leveling off yet. There are still significant quantities of halons in fire protection installations and halon banks; it appears that much is still being manufactured in some developing countries. So bromine will take longer to peak, before it starts to decrease. Methyl bromide, an important fumigant, is decomposed largely before it reaches the stratosphere but is still a concern. It does, however, appear that the methyl bromide concentrations are not increasing. There are also natural sources for methyl bromide, such as the oceans and biomass burning, and bromine compounds of human origin not controlled under the Montreal Protocol, such as ethyl dibromide. Overall, the Protocol is only addressing about one fourth of the bromine containing ozone-depleting compounds.

A few other issues merit a brief mention. Hydrochlorofluorocarbons (HCFCs), which are the replacements for CFCs, have been the subject of intense scrutiny by the atmospheric chemistry community and are now fairly well understood. One of their decomposition products is trifluoroacetic acid. However, HCFCs are not the only source of this compound which does not appear to be highly toxic. There is also some indication of liver damage caused by very high exposures to some HCFCs, but the levels shown to cause damage are much higher than any allowable exposure level.

In closing I would like to make a couple of comments on the connections between the Montreal Protocol and science. As I mentioned at the beginning, the scientific community was able to come to strong and convincing conclusions. One of the reasons for this development lies in the close connection between basic physics and chemistry and the issues that we are talking about. The scientific community has been able to describe and reproduce in the laboratory many of the chemical reactions taking place between a multitude of species found in the atmosphere. This is something that has not been possible with most other environmental issues. There are, however, many uncertainties remaining in the stratospheric ozone issue. For example, the biological effects are not nearly as well understood as we would like to; nevertheless, society has decided to take action, despite these uncertainties, because it is what prudence dictates.

It is uncertain how the ozone layer will behave in the future because the composition of the atmosphere is changing. There is now more carbon dioxide than in the past, and there will be even more in the future. The carbon dioxide will actually cool the stratosphere. Some other related questions remain: in what way will the addition of water by a fleet of supersonic airplanes impact the formation of clouds in the stratosphere? We have a number of interesting questions that need to be addressed from the scientific and regulatory perspective in order to insure the recovery of the ozone layer.

Technology brought about many advances, but we know that some of these advances have effects on the environment. The atmosphere is very fragile, and it is changing as the result of human activities. Some of these changes are damaging and should be avoided. An important lesson that springs from the Montreal Protocol is that these issues can be addressed if scientific, political, and environmental interests work together. We face some very important challenges in the next century: greenhouse gas and tropospheric ozone increases are two important examples. Fortunately, we have a precedent in the Montreal Protocol which shows us that these types of problems can be dealt with and solved.

THE STORY OF THE OZONE LAYER: LESSONS LEARNED AND IMPACTS ON THE FUTURE

Mostafa K. Tolba

On September the thirteenth, 1987, we did not have a Protocol; two days later, on the evening of the fifteenth, we had the Protocol. These were outstanding moments that marked the first major success in dealing with a global environment problem.

In his presentation, my dear Friend Professor Mario Molina has just addressed, with his usual and well-recognized authority, the scientific facts about the ozone layer and its depletion through human activities.

My remarks will concentrate on two points: innovation in the Montreal Protocol, and lessons learned from the negotiations of the Protocol. Naturally, I am basing my observations on my experience between 1984 and 1992 during which I was involved in the negotiations of the Protocol.

Innovation in the Montreal Protocol

The Montreal Protocol applied all known techniques and a series of innovative, sometimes revolutionary, approaches in order to facilitate agreement and ease its adaptation to evolving developments in science and technology.

The Protocol provided that decisions to adjust control measures regarding substances controlled by the Montreal Protocol become binding on all Parties to the Protocol when approved by a two-thirds majority, with no resort to the lengthy process of ratification.

Diplomatic conferences are, all too often, not in a position to deal with all relevant questions relating to an adopted instrument. Relatively minor, but nevertheless important, issues are frequently excluded from an agreed "package deal" on the understanding that relevant negotiations will immediately ensue. Two such issues during the negotiations of the Montreal Protocol were the confidentiality of data and the definitions of terms. Thus, a resolution of the diplomatic conference that adopted the Montreal Protocol requested the Executive Director of the United Nations Environment Programme (UNEP) to convene a working group to prepare a report clarifying data reporting requirements. Based on this report, the First Meeting of the Parties to the Montreal Protocol adopted decisions on the confidentiality of data and on the definitions of terms.

At the request of the former Soviet Union, the Montreal Protocol, under certain conditions, permits an increase of production beyond the agreed limits of the 1986 levels. At its first meeting, the conference *decided* by consensus, rather than through the lengthy process of amendment to the Protocol, that such a production increase may not be used for export to non-Parties to the Protocol.

At the famous Second Meeting of the Parties to the Montreal Protocol convened in London in 1990, states adopted a comprehensive amendment to the Protocol. Formally, amendments to protocols enter into force after at least two-thirds of the Parties of the protocol have submitted their instruments of ratification, acceptance or approval. Yet, the London Amendment required only twenty such instruments to enter into force. Some international lawyers go as far as considering that since this approach is not sanctioned by the language of the Convention, states may have tacitly amended the relevant provision of the Convention.

Annex A to the Protocol called for the ozone-depleting potential (ODP) of one of the controlled halons *to be determined*. The first Meeting of the Parties decided to accept the value for the ODP for halon 2402 as [6, 0] and to request the Secretariat to inform the depository that the Parties agreed to accept this figure by consensus and that, accordingly, the depository should insert this figure to replace the words "to be determined". This was done neither according to the amendment procedure provided in the Vienna Convention nor any specific provision of the Protocol. Its basis was the consensus among the Parties to the Protocol.

By a simple decision, the second Meeting of the Parties to the Montreal Protocol in London, established an Interim Multilateral Fund to support developing countries. Neither the Framework Convention nor the Protocol contains a specific legal basis for such a far-reaching step. Negotiations did not pay much attention to this issue at the time the Protocol was adopted in 1987. Politically, an early establishment of the funding mechanism was essential to induce developing countries to join the Protocol and accept its obligations. Several international lawyers consider that the establishment of a multimillion-dollar fund simply by decision of an intergovernmental body is an internationally unprecedented event.

Article 8 of the Montreal Protocol stipulates that "The Parties at their first meeting, shall consider and approve procedures and institutional mechanisms for determining non-compliance with the provisions of the Protocol and for the treatment of Parties found to be in non-compliance." This was adopted in 1987 and entered into force in January 1989. It took almost four more years before the Conference of the Parties adopted non-compliance procedures. Yet, these procedures lacked any indication of what should be considered as non-compliance, although they contained an indicative list of the measures the Parties might take against those who do not comply. However, the non-compliance procedures are considered by several international lawyers as unprecedented in international environmental law. So, in spite of its weakness, another first for the Montreal Protocol.

The Lessons Learned

We know the saga of Ozone is not yet over. But the ozone negotiations, between 1982 and 1992, have launched and established a new type of diplomacy, which can rightly be called global environmental diplomacy. What did we learn from these negotiations? Looking back at these negotiations, one realizes that there were a number of key ingredients which made a difference time and again.

The presence of a core group of countries with a similar objective. Starting before Montreal, a key group of countries was intent on moving towards a Protocol dealing with chlorofluorocarbons (CFCs), including the CANZ countries (Canada, Australia and New Zealand), the Nordic countries (Sweden, Norway, Denmark, Finland), the other European Free Trade Association (EFTA) countries (Austria and Switzerland), and the U.S. Without a continual push from those countries, I doubt that the Protocol would have been signed in Montreal. During the Montreal to London period, the most difficult issue was the establishment of the Fund. And here again, there was a core group of countries that played a major role (Mexico, India, China, the Netherlands, the Nordics, and the U.S. (save for that one difficult period when the White House played hardball). From London to Copenhagen, the core group on the Fund prevailed again. The control issues were relatively straightforward, except that we did not have a core group of countries on methyl bromide, and up to the end of 1992—the time I left UNEP—we did not achieve much on that issue.

The role of science and technology. I think it is fair to say that science and the consensus among scientists around the world were critical ingredients in the Protocol process, as was technology as well as the consensus that emerged on what could be accomplished and by when. Even more important was the assessment and reassessment process in the Protocol, since this forced a review and was almost impossible for the Parties to discount. In fact, without the *assessments for both science and technology*, I doubt that we would have been able to move either to the London or the Copenhagen agreements.

Willingness to compromise. While it is certainly true that the Montreal Protocol involved many countries with strong views on what should be done, there was always a willingness to take one step at a time. If we could not agree on a phase-out for CFCs in 1987, we accepted fifty percent. If we could not agree to a methyl chloroform phase-out by 2000 in London, we accepted 2003. If we could not reach an agreement on a phase-down schedule for methyl bromide in Copenhagen, we agreed to freeze. Of course, countries agreed to the interim steps because they believed they could come back another time and make the step more stringent, which is what happened at every stage. But I think the spirit of compromise was critical to the success of the Protocol.

Mobilizing the public is a sine qua non in environmental negotiations. The slow-motion Vienna Convention was accompanied by very little interest from non-governmental organizations (NGOs) and the media, and hence from the public. But, when scientists reported the Antarctic ozone hole and, more importantly, that the impacts of ozone depletion would include increased risk of cataract and skin cancer and of reduced body immunity, the public, especially in the North, became worried. Through NGOs and citizen groups, it pressed for quick action by governments. In only six years, between December 1986 and December 1992, this mobilization led to the negotiation, adoption, and entry into force of the Protocol, along with two huge amendments and very significant adjustments of control measures. It had taken fourteen years (1974 to 1988) from the time the two famous scientists Sherwood Roland and Mario Molina signaled the role of CFCs in ozone depletion, to have a framework convention in place and in force. It took two years to negotiate, adopt, sign, ratify and get the Montreal Protocol into force. The difference is essentially science and public concern.

The sponsorship of the negotiation by the United Nations Environment Programme (UNEP), thereby moving away from the traditional noncommittal neutral attitude of international organizations, also made a difference. Rather than adopting a passive role, I, as Executive Director of UNEP, and my colleagues from the Secretariat, particularly Iwona Rummel-Bulska and Peter Usher, played the role of mediators and citizens of the world concerned like anybody else with the ozone problems and eager to help find common grounds. It is worth mentioning in this respect that this view seems to be shared by some of the presiding officers over these negotiations. As Ambassador Winfried Lang, the Chairman of the Negotiations of the Montreal Protocol, points out:

> ...international organizations have assumed a critical role in virtually all multilateral negotiating endeavors. Beyond the normal tasks of hosting and servicing the respective Conferences, these organizations have been strongly interested in carrying out whatever secretariat functions were to be performed under the new treaties. This vested interest was to some extent missing from the negotiations on climate change, which were serviced by an ad hoc Secretariat. Thus, despite some worthy efforts, the secretariat on the whole did not exert strong pressure for a specific outcome of the negotiation, a situation quite unlike the negotiations on Ozone depletion.

Under a footnote, Lang states that "ozone negotiations were serviced by UNEP and strongly influenced by UNEP's top executive officer."

Informal consultations. This is probably the most important lesson learned. When government representatives sit around the formal negotiating table they are not relaxed. They are either timid or aggressive. They sure want to score points. This goal does not change in the informal consultations but representatives are more open since they are not formally committing their governments. Above all they get to know one another better. They get to see vividly the genuine interest of others to reach common solutions. They gradually become friends working for a common cause even if it is from different angles. Informal consultations were a key ingredient in negotiating the treaties that followed Montreal and that were negotiated under the auspices of UNEP: the Basel Convention on the Control of Transboundary Movement of Hazardous Wastes and Their Disposal, the preparatory meetings for the Climate Change Convention, and the Biodiversity Convention; and the recipe proved successful in all cases.

The negotiations of the Montreal Protocol also confirmed our understanding that in environment *everything is connected to everything else*. The negotiations of the Montreal Protocol brought to the surface the close links between protecting the ozone layer and controlling climate change.

During these negotiations we also witnessed the *unnoticed move away from the rigid positions on absolute sovereignty*. During these years of negotiations, governments were never considering that they were letting go part of their sovereign rights by allowing other states to tell them what they should or should not do within their own borders. The issue of sovereignty was never there, simply because each negotiating state was interfering in the internal affairs of all other negotiating states. States implicitly accepted that the concept of co-operation replace the insistence on absolute sovereignty.

While this was noticeable, we also noted that in spite of all the goodwill that prevailed and the common concern for the ozone layer, *governments were not, and I believe are not yet, ready to accept the urgent need for clear non-compliance procedures and strong action against those who do not comply.* Up to this time, the non-compliance procedures for the Montreal Protocol are still weak.

The *principle of common but differentiated responsibility* to confront environmental problems has existed since the Stockholm Conference in 1972, but the Protocol and its amendments crystallize and develop this approach in a new way. Parties recognized the need for greater equity between developed countries and developing countries and incorporated this need into the regulatory framework. In light of historical production and consumption of ozone-depleting substances in developed countries, it was agreed that developing countries should be allowed greater flexibility for phasing out these substances in the future—a ten year grace period. In addition, concrete financial resources were built into the legal framework essentially paid by developed countries but in the management of which both developing countries and developed countries had equal voice. The establishment of the Multilateral Fund and the commitment to technology transfer were crucial to the acceptance by developing countries of the obligations of the Montreal Protocol. I believe that similar actions would enhance the implementation of other environmental treaties such as the Basel Convention, the Biodiversity Convention, the Climate Change Convention and the Desertification Convention.

The *role of industries and financial institutions* was extremely important in the success of the development and implementation of the Montreal Protocol. The major chemical industries announced policies to phase out production of CFCs as soon as safer alternatives were available. Some, early in the game, set a 1995 deadline for halting CFC production. These phase-out policies sent customers a strong message to seek alternatives and substitutes.

Some industry associations and individual companies have phased out the use of controlled ozone-depleting substances. Many industry associations were engaged in extensive education, training and public awareness programs (especially through such measures as the voluntary labeling of products as "ozone-friendly"). In addition, the major chemical manufacturers of CFCs and halons have pledged not to sell or license CFC- or halon-manufacturing technology to countries that are not parties to the Montreal Protocol.

Such commitment by business and industry could have a great impact on the implementation of the Climate Change and Biodiversity Conventions, the Basel Convention on hazardous wastes as well as on the conclusions of new treaties on the use of Prior Informed Consent in Chemical Trade and on Persistent Organic Pollutants.

And last but not least, the presence of some *strong personalities*. In the end, it all boils down to individuals and personalities, and the Montreal Protocol had more than its share of strong and effective ones. Could we have reached what we did without the patience and perseverance of Winfried Lang, the Chairman of the negotiations that led to the adoption of the Montreal Protocol, or the strong leadership of his successors, the Chairmen of the Contracting Parties Conferences especially those in London and Copenhagen, Chris Patten (U.K.) and Kamal Nath (India)? I very much doubt it. Would we have a Fund without Juan Antonio Mateos (Mexico) and Ilkka Ristimaki (Finland)? It was fortunate that the G77 and the developed countries had leadership that knew when and where to take a stand, and when and where to compromise. Could we have moved the phase-out dates without Per Bakken (Norway), Vic Buxton (Canada), Richard Benedick and Eileen Claussen (U.S.), Laurens Jan Brinkhorst (Netherlands)? Could we have achieved anything without the very strong positions taken at various stages by Bob Watson, Rumen Boshkov, Dan Albritton, Lambert Kuijpers, Jan van der Leun, Patrick Széll, Steve Lebabty, Steve Anderson, Bill Reily and several others to explain, insist, argue, draft and redraft, push and pull? I very much doubt it. So I am convinced that, above all, successful negotiations require strong and effective participants; and while I cannot speak here about the other Conventions in whose negotiations I was involved, my view is that all of them had strong players as well.

Ladies and Gentlemen, in concluding this lecture, I wish to say that the regime established for protecting the ozone layer represents an innovative and holistic approach: the legal framework works in tandem with the scientific framework. Whereas the Convention recognized a general commitment for further research and a possible need for future regulation, the Protocol established actual control measures when scientific research proved that they were necessary. Furthermore, the Convention and Protocol mandate periodic assessments on which future regulatory measures can be based. The regime thus created allows for continual assessment and regulation, followed by further assessment, which will ensure the efficacy of the regime well into the future. More importantly, the Protocol established its own financial mechanism and specific commitments to relevant technology transfer.

This is the story of the ozone layer as I saw it unfolding and what I learned from following it up. I am aware, as I said earlier, that the ozone saga is not yet over. We still have to deal with compliance, illegal traffic, new control measures, alternatives to CFCs and so on. Yet, I claim that as a result of these negotiations we realize, as I said earlier, that there are a number of key ingredients which make a difference time and again in these negotiations. I believe the lessons learned during these negotiations were quite useful in conducting further environmental negotiations and will, in my view, influence other negotiations to come.

PART 2

THE ROLE OF SCIENCE

THE EVOLVING UV CLIMATE

James B. Kerr

Introduction

The extent to which ozone has been decreasing over the past two decades has been reported by many researchers and the results have been summarized in periodic international scientific assessments on ozone depletion (e.g. WMO, 1991; 1994). These assessments have quantified the long-term trends in ozone as functions of latitude and time of year and have summarized findings which associate these changes to enhanced levels of stratospheric chlorine and bromine that are anthropogenic in origin. The long-term trend values are based on the results of analyses of stratospheric ozone measurements which have been made routinely on a global scale for nearly four decades.

The quantification of long-term changes in UV-B irradiance at the earth's surface is more difficult than detecting long-term changes in ozone, since surface UV irradiance depends on many variables. Although analysis of spectral data can distinguish between changes due to ozone depletion and changes due to other causes, it is only relatively recently (since the late 1980s) that instruments have been developed with the capability of carrying out routine field measurements of spectral UV-B radiation with the required accuracy and long-term stability. These relatively short records have been analysed to quantify the dependencies of UV-B radiation as functions of ozone and other variables.

Figure 1 is a map that shows sites that are known to be monitoring spectral UV radiation at the ground. Data from the sites depicted by solid symbols have been reported to the World Ozone and Ultraviolet Radiation Data Centre (WOUDC) in Toronto and provide the database for results reported here. The database includes data from twenty-one sites with about 100 station-years of records that represent a reasonable coverage of the globe over a wide range of latitude and observing conditions. When data from other sites (depicted by open symbols on the map) become available, a more thorough analysis will be possible.

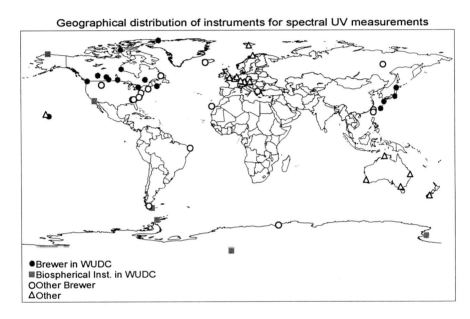

Figure 1. Location of spectral UV instruments

Dependence of UV Irradiance on Ozone

Figure 2 shows examples that illustrate how UV irradiance values at different wavelengths in the UV vary differently with total ozone. The measurements were made at Toronto for the period from 1989 to 1996 (left) and Montreal between 1993 and 1996 (right) when the solar zenith angle (ZA) is within 2.5° of 45°. Radiation at shorter wavelengths in the UV-B (300 nm) is absorbed strongly by ozone and shows a significant dependence on the amount of total ozone (top panels). Energy values increase by about a factor of 10 as ozone decreases from 450 to 250 Dobson Units (DU). The middle panels show that radiation at longer wavelengths in the UV-A (324 nm), where ozone has negligible absorption, has very little dependence on ozone. The bottom panels show that the UV Index (Kerr et al., 1994; Burrows et al., 1995), which is proportional to the erythemally weighted (McKinley and Diffey,

1987) irradiance at UV-B and UV-A wavelengths, is noticeably dependent on the amount of total ozone.

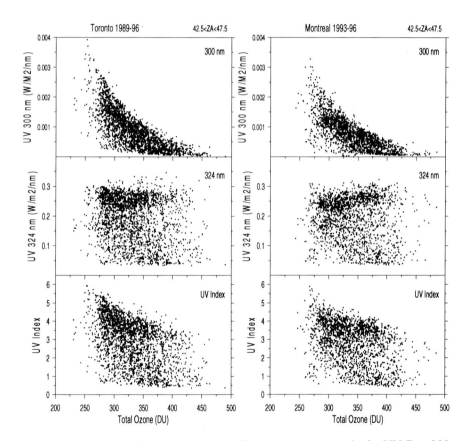

Figure 2. Influence of total ozone on irradiance measurements in the UV-B at 300 nm (top), in the UV-A at 324 nm (middle) and the UV Index (bottom) for Toronto (left) and Montreal (right). Measurements are for a solar zenith angle of about 45°. Results from Toronto and Montreal are very much the same.

Figure 2 illustrates that the dependence of irradiance values on total ozone as measured in Toronto is very much the same as that seen in Montreal. In fact the dependencies of UV irradiance on ozone are similar to the results shown above for the other stations reporting to the WOUDC.

These results illustrate the relationship between UV irradiance and ozone only for a zenith angle of about 45° and only at wavelengths of 300 nm and 324 nm. A more complete relationship over all wavelengths between 290 nm and 325 nm and for solar zenith angles between 25° and 90° has been quantified with a statistical model reported by Fioletov and Kerr (1997). Comparison of the observations with radiative transfer model is reported in Herman et al. (1996).

Dependence of UV Irradiance on Other Variables

Figure 2 illustrates that UV irradiance is significantly variable, even at a constant solar zenith angle (45°) and for a constant ozone value. The reason for the significant variability is due to other causes such as clouds, aerosols, surface reflection (albedo), and pollution. The data points in Figure 2 define fairly distinct upper limits which represent the clear sky conditions. The reduction in irradiance is due to scattering (by clouds and aerosols) and absorption (by pollution and absorbing aerosols) processes in the atmosphere. These processes can reduce the irradiance by more than 95% in some situations. These reductions are mostly due to clouds and aerosols and are in general about the same for all wavelengths (Kerr, 1997; Fioletov et al., 1997), except in some cases under very heavy convective clouds or when an UV absorbing gas (such as ozone or SO_2) is present in the atmosphere from sources such as pollution or a volcanic eruption.

A few isolated points lie above the upper limits in Figure 2 and are likely due to real causes such as snow on the ground which can enhance UV irradiance values by more than 30%, depending on location (Fioletov et al., 1997). In general, these enhancements are approximately the same at all wavelengths and can therefore be distinguished from ozone absorption which has strong wavelength dependence. Figure 3 shows an example from Resolute that clearly illustrates the enhancement in UV irradiance caused by snow. The data in Figure 3 are in the UV-A at 324 nm where there is negligible effect from ozone absorption. The diagram shows measured values as functions of time of day and time of year expressed as the percentage of the value expected under "clear sky" conditions. The clear sky values were determined statistically from the upper limit (see middle panels in Figure 2 for ZA = 45°) observed at all stations for all days without snow and for all zenith angles. From October to May, when there is snow on the ground and the sky is usually clear, most of the observations shown in Figure 3 are more than 100% of the clear sky value with up to 40% enhancement. From June to September, when there is usually no snow and the sky is often cloudy, the readings are lower and more variable.

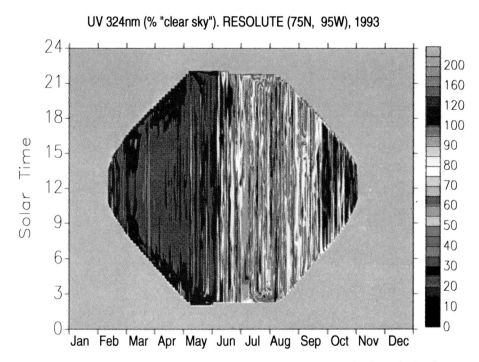

Figure 3. Measurements made at 324 nm relative to the expected "clear sky" value. Snow on the ground between September and May can enhance values by up to 40%. There is no influence of ozone since ozone has negligible absorption at 324 nm.

Present UV-B Climatology

There are now about 100 station-years of spectral UV irradiance data stored in the WOUDC. Many of the stations have records more than five years long. These data have been used to determine average UV values measured over the length of the records at each site. Examples of how these UV-B "climatologies" may be graphically summarized are shown in Figure 4. In these example, the average values of the observed UV Index are shown as a function of time of day and time of year at

Churchill, Canada (59N), San Diego (32N), Toronto (44N) and Sapporo, Japan (43N).

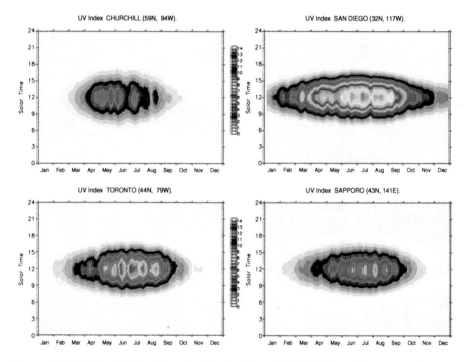

Figure 4. UV Index "climatology" plots for Churchill, San Diego, Toronto and Sapporo. Graphs show average values for several years.

Inspection of the diagrams in Figure 4 reveals several interesting behaviours of UV-B irradiance. At San Diego, Toronto and Sapporo the UV Index is lower near the spring equinox (March 21) than it is near the fall equinox (September 21). This is due to the fact that there is more ozone during spring than there is in fall. In Churchill, however, the UV Index is higher in spring than it is in the fall. This is because the UV levels are enhanced by snow on the ground which usually disappears in May or June. Comparison of the four station climatologies clearly shows higher levels of UV irradiance at lower latitudes. It also illustrates the longer hours of sunshine during summer and shorter hours during winter at higher latitudes. Comparison of the climatology at Toronto with that at Sapporo indicates that there can be significant differences in UV levels at two stations with similar latitude. These differences are due to different ozone and/or cloud conditions.

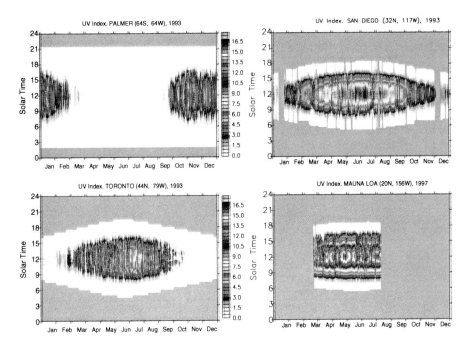

Figure 5. UV Index values for Palmer, Antarctica (1993), San Diego (1993), Toronto (1993) and Mauna Loa, Hawaii (1997). Palmer Station had higher values under the Antarctic ozone hole than San Diego did in summer. UV Index values at Mauna Loa are one of the highest observed (up to 16) under the low ozone values of the tropics with the sun nearly overhead and at high altitude (3500 m).

The measured climatologies that are evolving from the growing data base are useful to compare with short term variations. Figure 5 shows the UV Index for individual years at Palmer, Antarctica (64N, 1993), San Diego (32N, 1993), Toronto (43N, 1993) and Mauna Loa, Hawaii (20N, 1997). The observed day-to-day fluctuations are much larger than one might infer from the climatology diagrams shown in Figure 4. These fluctuations result in larger peak values than are seen in the climatological averages. For example, the climatology for Toronto shows a peak Index of less than 7 in June and July, but the individual observations made in 1993 (when ozone was a record low) show peak values greater than 9.5. From the perspective of biological effects, the episodes of extreme values on days with low ozone and clear skies may be just as important to consider as changes in the climatological averages. Figure 5 shows that Palmer under the ozone hole in October reads UV Index values greater than 11, suggesting very high sun burning potential. These values are more than those seen on a summer day in San Diego in southern California. The record at Mauna Loa illustrates that the combination of low ozone values in the tropics with the sun directly overhead and the high altitude (3500 m) results in UV Index values peaking near 16.

Estimating Past UV-B Climatology

In order to determine long-term changes in UV-B irradiance that may have occurred since the onset of ozone depletion in the late 1970s, it is necessary to estimate levels of UV-B irradiance that existed before routine spectral measurements were made. Estimates of past UV-B levels are the product of a thorough understanding of the dependencies of spectral UV-B irradiance on geophysical variables and applying these relationships with the aid of radiative transfer models to archived measurements of the other variables (e.g. ozone, cloud cover, optical depth, etc.). Work with the goal of reconstructing past UV-B climatology has been under way in recent years. These estimated past climatologies may then be compared with those of recent years to determine long-term changes.

Several studies have been done (e.g. Frederick et al., 1993; Madronich, 1992) that calculate changes in UV-B irradiance from long-term changes in total ozone measured by ground-based or satellite instruments. In deriving estimated changes in UV-B irradiance, most of these studies have assumed that the effects of clouds and other variables have not changed over time and are therefore representative of changes under clear sky conditions.

Kerr and McElroy (1993; 1994) reported results that quantify changes in spectral UV-B irradiation to changes in ozone under all types of weather conditions derived from measurements made at Toronto between 1989 and 1993. Applying the reported relationship between changes in ozone and irradiation to the known long-term trends in ozone (WMO, 1991; 1994), a long-term trend of 11% per decade would be determined for irradiation at 300 nm for summer months (May to August). This trend value has since been verified at Toronto through extension of the record both into the past (to 1986) and the future (to 1996).

More recently, Herman et al. (1996) have estimated past UV-B irradiance values at the surface from measurements made by the Total Ozone Mapping Satellite (TOMS). The TOMS instrument uses backscattered radiance measurements at several wavelengths in the UV to determine total ozone from space. The backscattered data may also be used to estimate cloud thickness and therefore derive the attenuation of surface UV-B irradiance due to clouds. Comparison of the TOMS estimated UV-B surface irradiance with actual ground-based irradiance measurements has been a key part of development of this approach (Eck et al., 1995). Although there are still problems to overcome with this method, the potential to provide a realistic estimation of surface UV-B irradiance values on a daily basis with global coverage back to the late 1970s is quite promising. The possibility to extend the record further back in time at sites where total ozone and cloud cover are measured also exists, although results of such analyses are yet to be reported.

Conclusion

Measurements of spectral UV irradiance are important to help our understanding of the behaviour of UV irradiance and to quantify relationships with the many variables that affect radiation at the earth's surface. Accurate replication of past and present global climatologies of UV is important for studies on the impacts of changing UV levels and can only be achieved through a better understanding of the relationships.

Spectral UV irradiance measurements have been made on a routine operational basis at network sites since the late 1980s and early 1990s and have proven to be quite reliable. The data-base for spectral UV irradiance measurements has been growing over the past few years and is expected to continue to grow in future years both with increasing length of the records as well as an increased number of stations measuring and reporting data.

Analyses of the data records have yielded information and results which were not known ten years ago. The analysis has matured from isolated studies that define results pertinent only to a specific site to more general studies that use data from a wide range of sites to define results on a global scale.

Acknowledgements

Dr. Vitali E. Fioletov carried out a significant part of the analysis reported here. Drs. D.I. Wardle and C.T. McElroy contributed fruitful scientific discussion.

References

Burrows, W., M. Vallee, D.I. Wardle, J.B. Kerr, L.J. Wilson and D.W. Tarasick. 1995. "The Canadian UV-B and Total Ozone Forecast Model." *Met. Apps.* 1: 247-265.

Burrows, W. 1997. "Cart Regression Models for Predicting UV Radiation at the Ground in the Presence of Cloud and Other Environmental Factors." *J. Appl. Meteorol.* 36: 531-544.

Eck, T.F., P.K. Bhartia and J.B. Kerr. 1995. "Satellite Estimation of Spectral UV-B Irradiance Using TOMS-Derived Total Ozone and UV Reflectivity." *Geophys. Res. Lett.* 22: 611-614.

Fioletov, V.E. and J.B. Kerr. 1997. "Numerical Relationship Between UV Irradiance, Total Ozone, and Other Variables from Analysis of Brewer Spectral UV-B Measurements Archived at the World Ozone and UV Data Centre." *Proceedings of the Quadrennial Ozone Symposium* (forthcoming).

Fioletov, V.E., J.B. Kerr and D.I. Wardle. 1997. "The Relationship Between Total Ozone and Spectral UV Irradiance from Brewer Spectrophotometer Observations and its Use for Derivation of Total Ozone from UV Measurements." Submitted to *GRL*.

Frederick, J.E., P.F. Soulen, S.B. Diaz, I. Smolskaia, C.R. Booth, T. Lucas and D. Neuschuler. 1993. "Solar Ultraviolet Irradiance Observed from Southern Argentina: September 1990 to March 1991." *J. Geophys. Res.* 98: 8891-8897.

Herman, J.R., P.K. Bhartia, J. Ziemke, Z. Ahmad and D. Larko. 1996. "UV-B Radiation Increases (1979–1992) from Decreases in Total Ozone." *Geophys. Res. Lett.* 23: 2117-2120.

Kerr, J.B., C.T. McElroy, D.I. Wardle and D.W. Tarasick. 1994. "The Canadian Ozone Watch and UV-B Advisory Programs." In *Ozone in the Troposphere and Stratosphere*. Proceedings of the Quadrennial Ozone Symposium 1992, NASA Conference Publication 3266, Berlin: Springer-Verlag, 794-797.

Kerr, J.B. and C.T. McElroy. 1993. "Evidence for Large Upward Trends of Ultraviolet-B Radiation Linked to Ozone Depletion." *Science.* 262: 1032-1034.

Kerr, J.B. and C.T. McElroy. 1994. "Analyzing Ultraviolet-B Radiation: Is There a Trend?" *Science.* 264: 1342-1343.

Kerr, J.B. 1997. "Observed Dependencies of Atmospheric UV Radiation and Trends." *NATO ASI Series.* Vol. I 52, Eds. C.S. Zerefos and A.F. Bais, Berlin: Springer-Verlag, 259-266.

Madronich S. 1992. "Implications of Recent Total Ozone Measurements for Biologically Active Altraviolet Radiation Reaching the Earth's Surface." *Geophys. Res. Lett.* 19: 37-40.

McKinley, A.F. and B.L. Diffey. 1987. "A Reference Spectrum for Ultraviolet Induced Erythema in Human Skin." In *Human Exposure to Ultraviolet Radiation: Risks and Regulations.* Edited by W.R. Passchler and B.F.M. Bosnajokovic, Amsterdam: Elsevier.

WMO. 1991. *Scientific Assessment of Stratospheric Ozone: 1991.* World Meteorological Organization. Global Ozone Research and Monitoring Project, Report No. 25. Geneva: World Meteorological Organization.

WMO. 1994. *Scientific Assessment of Stratospheric Ozone: 1994.* World Meteorological Organization. Global Ozone Research and Monitoring Project, Report No. 37. Geneva: World Meteorological Organization.

UV-B EFFECTS ON AQUATIC ECOSYSTEMS

Robert Worrest

I am going to go very briefly over a summary of the effects of UV-B radiation on aquatic systems. We have been studying this for many years now, and to summarize this in the time available will be somewhat difficult. I will just give you some highlights that will be coming out in reports that we have in preparation by the United Nations Environment Programme (UNEP) Environmental Effects Assessment Panel.

Phytoplankton and other organisms fix carbon dioxide in the presence of sunlight and produce carbohydrates, structural proteins, and organic molecules for use in both the plant and animal kingdom. In the aquatic ecosystem, it has been calculated that they produce between 90 and 100 gigatons of carbon per year. This is approximately equal to that carbon fixed by the terrestrial ecosystem.

About 99% of aquatic systems are made up of marine systems which is why the major emphasis of studies of UV-B effects is on the marine ecosystems. Within the marine ecosystem the phytoplankton, the small carbon dioxide fixing organisms which form the base of the marine food web, are by far the most important component. Any reduction in carbon uptake as a result of factors such as UV-B impact upon phytoplankton, could result in increased carbon dioxide in the atmosphere and therefore an increase in the greenhouse effect.

There are several questions that need to be discussed when we talk about the impacts of ultraviolet radiation on marine ecosystems, or on any aquatic ecosystems. Five questions that we have been working on, some of them being answered by atmospheric scientists, are:
1) What are the predicted changes in ozone concentration and UV-B radiation on a global basis during the decades to come? This question is to determine what the primary impact of UV-B radiation is on these systems.

2) What is the spectral penetration of this ultraviolet radiation in the aquatic system, such as in the oceans? This varies depending upon the concentration of various filtering substances within the aquatic systems.
3) What is the vertical distribution of the aquatic organisms in the water column? Are these sensitive organisms near the top of the water column or are they mixed? This plays a role in the various detrimental effects of ultraviolet radiation on the system because if they get mixed around they will be near the surface at some times, and farther down in the water column at others where they potentially have greater protection.
4) What is the spectral sensitivity of the organisms? This varies from organism to organism and also for organic compounds or molecular substances within the organisms. The sensitivity varies from substance to substance, making it very difficult to try to determine what the impact of any potential increase in ultraviolet radiation might be, because of the variability of sensitivity of organisms within species, between species, and within molecules within the organism itself.
5) What are the extent and limits of UV repair and adaptation? What are some of the other factors that play a role?

These are all areas of research that are ongoing. The UNEP assessment reports of 1991 and 1994 provided information on the state of knowledge at that time and will be updated in the new assessment scheduled to appear in late 1998.

The distribution of organisms within the marine system is quite patchy. We have a non-uniform distribution of phytoplankton within the marine systems with a very predominant concentration at the higher latitudes and in the upwelling regions of the temperate latitudes. There are various factors that determine this lack of uniformity such as nutrient concentrations—these organisms require nutrients such as nitrates and phosphates in order to grow—light availability, and water column stability. There is speculation that UV-B radiation plays a role in limiting the development or growth of organisms within the tropical regions, along with nutrient limitations.

Satellite technology is giving us new insights into the distribution of phytoplankton. Although coastal zone color scanner data goes back to 1978, a recent much delayed launch of a satellite sensor is now starting, from the beginning of this month (September 1997), to give us much more information on the distribution of phytoplankton within the oceanic systems.

The penetration of solar radiation into the water column does vary. It is controlled not only by suspended substances or particulate material, but also by dissolved organic material. The concentrations of these filtering substances within the water column varies. You get relatively high concentrations near land and areas of high productivity, and lower concentrations elsewhere. So there is a huge variation in the degree of penetration of UV-B radiation into the water column and therefore, a tremendous variation in the potential impact upon the life forms within that water column.

An area of considerable research interest is the impact of solar ultraviolet radiation upon blue-green algae, a nitrogen fixer. It is of considerable importance to agriculture. Huge areas of rice paddies are seeded with blue-green algae to fix nitrogen from the atmosphere into nitrates and nitrites within the water column. Higher plants and many phytoplankton cannot fix nitrogen, and depend upon a source of nitrates, one of which is blue-green algae. These algae extract nitrogen from the atmosphere and create nitrate; thereby creating a fertilizer from the atmosphere. The amount of nitrate being produced is in the order of three times as large as the amount of nitrogen fertilizer being synthesized by humans from ammonia.

So you get a tremendous increase in the amount of available nitrate because of these nitrogen-fixing organisms. It turns out that the enzymes for nitrogen fixation are extremely sensitive to UV-B exposure. Also the motility, orientation, the generation of photosynthetic pigments and oxygen production of these nitrogen-fixing organisms are highly UV-B sensitive. So this is one class of organisms that has been shown to be of concern in the face of potentially increased levels of solar ultraviolet radiation.

Many organisms can actually produce screening pigments. Sensitive organisms, such as DNA, are protected in response to UV radiation exposure by the generation of these screening pigments. Its very analogous to the human production of melanin pigments which protect us from damaging ultraviolet in response to preliminary exposure to UV radiation. Balancing off some of these damaging impacts are the effects of longer wavelengths UV and visible radiation that can help repair some of the damage or protect against UV radiation damage.

I showed earlier that the oceans are a major sink for atmospheric carbon dioxide, not only through carbon dioxide dissolving into the aquatic medium, but also by being taken up by the phytoplankton and other algal components within the marine systems. In the marine food web, you have very intricate networks of systems, and dependence from one system to another, one population to another. This includes bacteria, phytoplankton, small animals and zooplankton which are fed upon by other organisms. In the Antarctic, you have a very short food chain between the phytoplankton, krill, and organisms such as the whales. But in other areas of the marine system this food web is far more complex. If we consider the total protein consumption around the world, it has been shown that about 30% of the world's animal protein for human consumption comes from the sea. About 15% comes from fish alone. In many developing countries this percentage is even larger than 30%. It is very difficult to carry out research activities in these areas. To try to simulate the natural marine system is an expensive process and therefore there has been very little research. Preliminary work occurred back in the 1970s and early 1980s but very little has occurred since.

Regarding consequences, we have shown that current levels of exposure to UV radiation are stressful to many freshwater and marine organisms. Laboratory ecosystem-simulating experiments show that enhanced levels of UV radiation also have an increased impact upon these organisms. Quantitative estimates on the total ecosystem are very difficult. It is hard to extrapolate to a total ecosystem from individual studies on organisms. Many of the ecosystem studies in large flow-through tanks, although they do simulate systems, are still quite artificial in trying to simulate the activities going on within natural ecosystems.

If we do have a reduction in biomass production it would also, by implication, have an impact on the food supply for those countries and populations that depend upon marine sources for their protein intake. If you have a change in species composition because of the variation in the sensitivity of the various organisms, you can change the ecosystem makeup, and therefore have an impact on the total ecosystem in that particular area. Decreased nitrogen stimulation by nitrogen-fixing organisms, such as the cyanobacteria or other prokaryotes, could cause a nitrogen deficiency in those areas of agriculture that depend upon these nitrogen-fixing organisms for their nitrogen fertilizer.

Finally, I would just note that a decrease in marine primary productivity would result in a decrease in the carbon dioxide uptake and, therefore, would have an impact on the global climate issue and global warming.

EFFECTS OF UV-B ON PLANTS AND TERRESTRIAL ECOSYSTEMS

Manfred Tevini

Thank you very much for inviting me to speak here. I would like to address several questions that have been important during the last ten years, the most important of which is: What are the effects, if any, of higher levels of UV-B on terrestrial ecosystems?

We are all familiar with effects of human overexposure to the sun. But the picture of ultraviolet radiation damage is not as simple as you might think. In the case of sunburn the exposure to the sun may be too long, or the intensity too high. The product of both is a daily dose which is increased by ozone depletion. A third factor, which is sometimes overlooked, is that there might be too much power in the ultraviolet radiation. A shift to shorter wavelengths means that there is more energy in the photon with higher potential to damage biological molecules. All three can happen with ozone reduction.

Can that happen to plants too? It is easy to very quickly knock out plants if you use wavelengths which are too short, those that are not present in our natural environment today but have been present during prehistoric times. So in investigating plant impacts it is very important to use the right experimental design with UV light sources which are adapted to natural sunlight. This was not true in the early experiments.

Why is ultraviolet radiation so dangerous to plants? There is absorption by very important molecules present in every organism, such as the genetic material DNA which can be damaged by very powerful photons. That means that information contained in the DNA cannot be passed from cell to cell, and plant damage results. Another compound, called IAA, a specific plant growth regulator, interacts with many growth responses, for instance with shoot elongation and/or flowering for which other UV-absorbing phytohormones "the gibberellines" are responsible. If these regulators are impacted you can also expect damage or developmental changes in the plant.

What is the approach to studying enhanced UV radiation? It is not that easy. We cannot just take plants from Montreal and move them to Florida because there is a natural higher UV radiation nearer the equator. There are also different climatic conditions so it is not just a situation of enhanced UV radiation. To do good studies we have to change UV radiation *in situ*, but that too poses problems.

So what has been done so far? The scientific community started experiments mainly with artificial radiation in growth chambers using UV-B tubes and white light tubes, or in growth chambers using high intensity lamps. These were not very satisfactory as you need a careful balance of visible, UV-A and UV-B radiation. You can in fact repair DNA with higher amounts of UV-A, blue light and white light, so simulating the right balance is important. Using xenon lamps, which have a much higher intensity, is a better approach. You can also use artificial light sources in greenhouses, but these have shadows, meaning reductions in white light, so it is not exactly what you would expect out in the field. We then went on to artificial UV-B together with natural solar radiation in the field, either by supplementing UV-B using lamps that were either on or off, or by supplementing the UV-B by increasing it according to a daily course. That is the most sophisticated equipment that we have nowadays.

Because of criticisms regarding artificial radiation, work has also been done using only solar radiation. This can be done in growth chambers covered by a cuvette containing a stream of ozone. In this way the solar radiation has to pass through ozone, which is the natural UV absorber in the stratosphere, and the radiation entering the growth chamber is reduced by a desired amount. So you have a perfect simulation of reduction of UV rather than enhancement. You can take this reduced UV case as a control and look upon the natural case as relative enhancement. This approach has been used in Portugal, where the ambient radiation is very high. One thing this type of experiment has shown is that even under natural conditions there are UV-B impacts on plants by relatively enhanced UV-B.

Another approach, also tried in Portugal, is to use greenhouses that are very transparent, much more so than normal greenhouses. By using Plexiglas of different thickness, one of 3 mm and another of 6 mm, you can simulate a relative UV enhancement of 10 to 15%. Because this uses only natural solar radiation it is a considerable advantage.

Is there any general impact on all plant species? The answer is no. Ultraviolet impacts depend on species and cultivars, and also on growth and climatic condition. Alan Teramura showed with nineteen cultivars of soybeans that over 80% raised in growth chambers are sensitive as shown by reduction in biomass. In greenhouses, there is a reduced sensitivity, and in the field only about one-fifth are sensitive. That clearly shows that there is an overestimation of effect if you take only growth chamber or greenhouse results because the repair mechanisms, which are stimulated by visible sunlight, are suppressed.

What specific UV-sensitivities have been found so far? In various species and cultivars we see changes in plant height, leaf length, and leaf area. Leaf thickness is increased, branching is increased, leaf and plant architecture is altered, and canopy structure is altered. Very importantly, timing of development and flowering is delayed in some cases. Even seed production is delayed in sensitive species such as soybeans, rice and green beans. There is a lot of sensitivity between species and cultivars, especially under combined stress conditions where the sensitivity can increase or decrease. That is all the result of defense mechanisms that protect the cellular mesophyll inside a leaf tissue, or repair the DNA genetic material. The very sensitive rice seedling cultivar Fanion shows a large impact on stem growth with a decrease of 1% biomass at an equivalent ozone depletion of 1%. So this is a one to one relationship, but an extreme case seldom found.

There is a famous experiment done by Alan Teramura who showed that the forest species, loblolly pine, is reduced in length, with an accompanying change in plant structure, when ozone is reduced by 25 to 40%. Another finding relates to suppression of flowering. The species *Hyoscyamus niger* only flowers when the day length is sufficiently long. However, if you take long days and enhance UV by only a small amount you cannot induce the plant to flower.

Overall for agro-ecosystems, in particular for sensitive species, we find changes in timing of flowering, reduction in growth, and changes in growth form. These may change the competitive balance between weed and agricultural species. Is there an impact on food yield? So far we do not have much quantitative data but we do suspect that both food quantity (yield) and quality may be impacted. For yield loss, one study with the soybean cultivar Essex showed a 25% loss with 25% ozone reduction. In Portugal, even when we simulated only a small relative ozone reduction, we found a potential yield loss of 20% with green beans. However, there are several experiments that have not found any changes in yield comparable to reductions in biomass. Experiments in the Philippines and Japan taken together showed in rice plants biomass reduction of 40 to 50% but a yield reduction of only 25%. In Portugal, in an experiment with eight maize cultivars, simulating 10-15% higher UV-B radiation we found a yield reduction of about 20%. However, that was at the first harvest. There was a delay in cob development and not all cobs were ripe when harvested. Harvesting adjacent plots two to three weeks later, we found no significant reductions in yield.

Is there any change in food quality due to higher UV-B radiation? Bioactive substances such as flavonoids shielding the plants may act as protective antioxidants. That can be a positive effect for human health. In spice plants we found an increase in etheric oils, which makes basil much more tasty than under normal conditions. There is also an impact on proteins as shown in soybeans. In Alan Teramura's six-year experiment with the soybean Essex which showed a 25% reduction in yield, in two of the years there was a yield increase. He attributes that to the stimulation of protective mechanisms for the plant because of drought. The chemicals involved in that protection could impact on food quality. It is interesting to consider the agricultural merits of a plant that is UV sensitive but more resistant to other stresses.

As I mentioned, quantities of protective pigments are increasing. But not only are the flavonoids increasing but there are also some compounds of consumer interest. Ascorbic acid in melons shows an increase of three times or so, and also sugar content is significantly increased under enhanced UV-B. That was not expected. So, we also have positive effects in different plant species when UV-B is increasing in the future.

Can natural ecosystems as well as agro-ecosystems be impacted by enhanced UV-B? In general, plant responses are the same as for agricultural cultivars. There is growth reduction, for example, in cranberries in Northern Sweden grown under about 25% ozone depletion scenario. Growth form also changed as well as developmental timing. One might expect that if there is any time delay, the natural pollinators would not necessarily be available and pollination would be affected. The reproduction of that species would then be hindered and the competitive balance and species composition in that ecosystem may be changed. Biodiversity is one of the most important things we should look at in the next five or ten years. Let us hope that funding will be available not only for five years, but for a hundred years.

Secondary chemistry, as with agricultural species, is also variable in natural ecosystems under higher UV. That has indirect effects on herbivores because some of these compounds are fungicidal or bactericidal. A change in secondary compounds has thus indirect effects on insect or pathogen attacks. By changing secondary compounds, like flavonoids and lignins, different decomposition down in the soil can be expected. Even the microorganisms in the soil are impacted by the secondary compounds in a way that we do not yet know anything about. Neither do we know much about effects on nutrient cycling, which comes back to the effect on competitive balance and the biogeochemical cycle.

In closing, I would like to address some current needs for assessing UV impact on terrestrial vegetation. Effects of modest ozone depletion are not clear, say changes below 15% under field condition. We have very few studies on food quantity. Research should be increased for both food quantity and quality. We do not have much information on the interaction between phytochemical changes, pathogens and insects, and on reasons for tolerance. There may be scope for genetic engineering here because if we know the reasons for sensitivity then we can, by inducing genes which increase protective pigments or repair mechanisms, breed new UV-resistant cultivars. There is only a little information about natural communities, much less than for other air pollutants, elevated CO_2, or temperature. And we have very few results about the overall impact on biogeochemical cycles.

However, despite all the things we do not know, we do have enough information about negative effects on plant ecosystems by enhanced UV-B to know that we must go on protecting our earth, especially the ozone layer.

STRATOSPHERIC OZONE DEPLETION AND UV-INDUCED IMMUNE SUPPRESSION: IMPLICATIONS FOR HUMAN HEALTH

Edward C. De Fabo

First of all I would like to thank the organizers, and in particular Environment Canada and the United Nations Environment Programme (UNEP), for the invitation to speak to you today on the topic I have been associated with for the past twenty years, the effects of increased solar ultraviolet B radiation (UV-B), due to stratospheric ozone depletion, on biological and health systems.

As a research scientist working on UV-B induced immune suppression, and having been the chair of two international science advisory groups on UV effects on the biosphere, I have really come to appreciate the role and sometimes dramatic effects that UV-B radiation can have on living cells and systems. Today, I would like to share some of this information with you.

By way of definition, UV-B is that radiation found in the 290 to 320 nanometer waveband, and these are the shortest rays of non-ionizing sunlight that reach the earth's surface. The fact that UV-B can perturb specific biological systems has been well known for many years. Mutagenic changes to DNA and denaturing or damaging changes to proteins are examples of damage to two of the most important biomolecules which are common to all living cells. However, what is not so well known are the overall UV-B induced changes to homeostatic mechanisms controlling human health and ecosystems. It is worthwhile noting that while human health tends to get the most attention, the fact is there is an indispensable interrelationship between all members of the biosphere. What happens to seemingly esoteric ecosystems in one part of the world can affect ecosystems in other parts of the world, and ultimately affect human health and welfare. It is for this reason that global ozone depletion represents one of the most serious environmental problems which the biosphere has ever faced.

I would like to begin with a very troublesome piece of information. If one looks at the lifetimes of chlorofluorocarbons (CFCs) in the atmosphere, one can see that it will take something like fifty or sixty years to achieve pre-1980 levels of ozone, and this will come about only if *all* countries comply with the Montreal Protocol. Thus the length of time humankind is expected to live with higher than normal UV-B solar radiation, over and above natural fluctuation is, therefore, on the order of fifty to sixty years or even longer.

Is there a need for heightened concern? Over the past two years, I have been collecting data from Environment Canada on stratospheric ozone levels over eight Canadian cities. Compared to the pre-1980 levels there are far more weeks of negative ozone departure than weeks of positive ozone departure, something on the order of six to one. See for example figure 1 showing losses over Montreal for the past two years or so. Across all Canadian cities monitored, there is an overall average of about 3 or 4% decrease, with a lot of episodic fluctuation superimposed, sometimes reaching as high as 20%. Thankfully, the Montreal Protocol, although not a final cure, has made a significant step forward since without it these trends would be even more disconcerting.

Nonetheless, specific, accurate, quantitative predictions of most of the effects of increased UV-B on human health over long or even short time periods cannot effectively be made at present because of the severe lack of detailed data essential to such predictions. Some risk assessment is possible, and Dr. van der Leun in this volume, addresses this aspect with respect to skin cancer. We do know, however, that as ozone levels decrease solar UV-B radiation increases, generally on the order of 1 to 2% for each 1% decrease in ozone. Above about 20% depletion, UV-B increases can become exponential as we know from Dr. Kerr's contribution. I am stressing the UV-B as this is the portion of the solar spectrum that increases with stratospheric ozone decreases.

A big question surrounding ozone depletion is how much of a UV-B increase is needed to begin observing effects? It is a difficult question to answer. For example, some of the factors which can effect UV-B reaching a biological target are pollution, clouds, rain, smog, clothing, skin cells and pigmentation. The point is that in order for the UV-B photon to do any damage it has to be absorbed by the target. I should also point out that only recently have ground-based measurements of UV radiation, both on a broadband and spectral basis, been established in different countries around the world. To its credit, Environment Canada has been in the forefront of daily UV radiation and ozone monitoring for many years. In the United States, up until a few years ago, attempts at measuring UV radiation were limited to a few selected sites and with instruments that were not all that sensitive to begin with. We in the United States are still behind Canada and Europe, but serious attempts to measure ground-based UV-B on a nationally coordinated scale appears to be developing. That is encouraging, but movement is painfully slow. Thus, our information about how much UV radiation has been received at the earth's surface, either globally or locally, is rather limited in terms of making risk assessments, particularly in the health areas with which I am involved.

What are those health areas most likely affected by ozone depletion? Skin cancer comes first to mind and Dr. van der Leun will talk about this subject so I will only

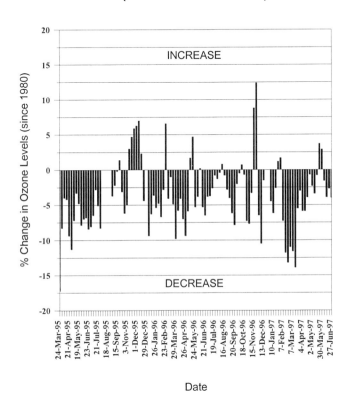

Figure 1. Ozone Level Changes in Montreal

provide a brief introduction to this disease by way of immune suppression because the two are linked. Eye damage, certain infectious diseases, and immune suppression are other health areas likely to be affected by ozone depletion. I would like to emphasize *certain* infectious diseases because not all infectious diseases will be influenced. In fact, we do not have very much information on links between infectious diseases, ozone depletion and UV-B radiation at all. Based on very limited information, certain infectious diseases such Leishmaniasis and Herpes simplex, may be affected by the impact of UV-B on the immune system but much more research is needed. There is data on a few other experimental systems, but virtually no human data. Interestingly, recent new studies appear to implicate UV radiation and immune suppression with non-infectious diseases such as Multiple Sclerosis and Non-Hodgkins lymphoma. Though few in number, studies such as these clearly bear further investigation.

I would like to say a very brief word about cataract, even though this is not involved directly with immune suppression. This is another serious health problem that has to be considered. UV-B radiation has been implicated in some cataract formation but no clear cut answers are currently available. In my own opinion, increase in cataracts will be one of the major health impacts of increased UV-B radiation over extended periods of time.

The area with which I am most familiar is immune suppression by ultraviolet radiation. My wife and colleague Dr. Frances Noonan and I have been involved in this area of research since the late 1970s. Historically the experiments of Margaret Kripke and Michael Fisher were the first to show the dramatic effects that UV-B radiation had on a mouse's immune system. Basically, UV-B down-regulates the cell-mediated arm of the immune system preventing it from attacking skin tumors. It had been well known for many years that irradiating mice with a sun lamp, usually for about one year, could produce skin tumors. Fisher and Kripke noted that if one tried to transplant such a tumor into a genetically identical mouse the tumor would be rejected. The seminal observation these investigators made was to show that pre-irradiation of a mouse with the same UV sun lamp before transplanting the tumor resulted in successful transplant of that tumor. In short, the same sun lamp which could induce tumors over a long time period could also affect the mouse immune system's ability to destroy that tumor. Shortly after these initial experiments were completed, I joined the group at Frederick Cancer Research Center and began to ask some photobiological questions such as: What are the wavelengths within the broadband emitting sun lamps that are responsible for suppression of tumor rejection? What was the shape of the UV dose response? Is suppression of tumor rejection a function of the total dose and thus independent of dose rate and time? And so on. We also asked whether there were other immune responses which might show suppressive effects by UV. The answers to these questions have been published in the scientific literature for some time now.

The methodology we use in studying UV impacts is the contact hypersensitivity or CHS response. This response is central to our studies and has remained the core end point of most of our work since the late 1970s. Essentially, it works like this. First, the shaved dorsal or back side of the mouse is irradiated with UV-B. A contact allergen such as trinitrochlorobenzene is applied either to the UV exposed site or a non-UV exposed site. Several days later, one elicits an immune response by applying the same contact sensitizer to both sides of the ear and measures the immune response which is seen as an increase in ear swelling. This CHS response measures the function of a major class of T-lymphocytes. The critical finding we made was that UV radiation can down-regulate the CHS response as a function of wavelength and dose. Over the next several years, we studied in detail UV suppression of tumor rejection, and UV suppression of contact hypersensitivity. It should be noted that, at the moment, the antibody arm of the immune system appears not to be affected by UV-B.

Summarizing many years work in a few words, we found that, because of the unexpected similarity of the photoimmunological parameters between the two systems, the most striking conclusion we could draw was that there appeared to be a common step between these very different responses. We reasoned that this common step was most likely occurring at the level of the initial absorption of UV-B. This suggested the existence of an unknown UV absorber in skin, a photoreceptor, which could absorb UV-B and convert the radiant energy of the photon into a biochemical signal which would then set in motion a series of biochemical events leading to down-regulation of certain cell-mediated immune response.

The question then became how to find a putative photoreceptor in skin among so many UV absorbing compounds present. Our task was akin to looking for the proverbial needle in a haystack. Given that radiation has to be absorbed before anything can happen, and realizing, as stated, that the skin has plenty of UV absorbing substances, the objective of our study was to identify a specific substance which could mediate interaction between sunlight and the immune system. Basically, it meant that we needed to determine the identity of the substance in skin which could, in fact, act like an antenna on a TV set. That is, receive incoming electromagnetic energy signals, in our case UV-B, and convert this radiant energy into a biochemical signal which could communicate with the spleen and lymph nodes, the immune organs where UV suppressor signals were known to be found.

To appreciate the difficult nature of this suggestion, allow me to use the following analogy. I take a laser pointer beam and shine it on the wall. Now I try to convince you that the beam is going through the wall into the next room. And not only that, but the energy in the beam, once it gets through the wall, is able to rearrange the furniture in the next room. It sounds rather crazy and, for a very long time, a lot of people thought it was. But if I tell you that there is a substance on the wall which could absorb the radiant energy of the laser beam and convert it to another energy form able to penetrate walls, it then begins to make sense. All the pre-existing experimental data that we had accumulated at that time was pointing to the existence of such a unique immune-regulating photoreceptor, or signal transducer, capable of converting UV-B energy into a biochemical signal capable of down-modulating cell-mediated immunity. This idea was postulated without benefit of any precedent in the literature like it.

How can one go about trying to find such a compound and prove its existence? We used an experimental technique that allowed us to separate the active waveband of immune suppression, the UV-B, into very narrow bands or slices of wavelengths within the UV-B region by the use of interference filters coupled to a 2,500 watt xenon arc. By producing these narrow wavelengths with half-bandwidth of 2.5 nanometers across a relatively large area, 50-60 square centimeters, we were able to irradiate the shaved dorsal surface of three mice simultaneously with narrow band radiation in 5 nanometer steps from 250 to 320 nanometers. This was a first of its kind system for use across the ultraviolet spectrum. This system allowed us to determine individual dose-response curves, which in turn allowed us to construct a detailed spectrally well-defined in vivo *action* or wavelength dependent spectrum. In theory such an action spectrum should be exactly congruent to the in vivo absorption spectrum of the putative photoreceptor or compound absorbing the radiation. In short, we were defining, *in vivo*, its absorption spectrum. We did such an experiment. It took nearly two years to complete as we needed to irradiate 1500 mice, make 6000 ear measurements and measure UV irradiance before and after every experiment. When completed, the action spectrum showed a peak at 270 nanometers, a shoulder at about 280 nanometers, and then a slow decline to 320 nanometers. This matched quite nicely the in vivo absorption spectrum of urocanic acid (UCA). There was another possibility for the photoreceptor which we considered, DNA. Pyrimadine dimers which are a major photo-product formed in DNA by UV radiation, is a proposed mechanism also for UV immune suppression. However, we found a serious lack of correspondence between our experimental action spectrum for immune suppression and the absorption spectrum for DNA.

UCA is a derivative of the essential amino acid histidine and is formed in the skin when histidine is de-aminated in the trans-isomeric form. Upon absorption of UV radiation, trans-UCA is able to isomerize to the cis isomer. It is the cis isomer which we propose is responsible for producing the immunosuppressive events. This, in its most simplistic form, is our working model: the reaction begins with the photoisomerization of trans-UCA, to the cis isomer. We hypothesize that cis-UCA either directly or indirectly interacts with a binding receptor and initiates a cascade of events leading to secondary signals. These signals then produce a modification to certain immune cells. How it does this is unclear but whatever the mechanism, the end result appears to be a modification or change in the processing of antigen or foreign substance by specialized immune cells known as antigen-presenting cells (APC). Specifically, we postulate that once cis-UCA modifies the APC, when it processes antigen, a down-modulating or suppressive signal is produced instead of the usual up-modulating or effector signal against the antigen. Depending on whether the processed antigen is chemical-, viral- or tumor- "associated", the suppressive signal appears to be specific only for that particular antigen. We believe this to be *the* fundamental reaction in the whole process.

Because of time I can only summarize for you what is known about immune suppression. We know that it occurs in humans, is independent of skin pigmentation, is directly linked to skin tumor outgrowth in experimental studies, and it may play a role in certain infectious diseases. The action spectrum for immune suppression shows significant activity in the UV-B, but we do not know much about the role of radiation beyond the UV-B. Interestingly, UV-A may have modulatory control over the UV-B immunosuppressive response. This is an area of active investigation in our lab at the moment. Further, UV-immunosuppression shows a radiation amplification factor of 0.6 to 0.9 depending on latitude, meaning that for each 1% decrease in ozone there is a 0.6–0.9% increase in biologically effective sunlight irradiance for immune suppression.

People sometimes ask why would nature evolve a system which would lead to the development of skin tumors. While the answer to this is not certain, UCA may be a natural protective mechanism designed to suppress auto-immune attack against sun-damaged skin. For example, sun-burned skin cells are severely damaged cells and might look "foreign" to the immune system. This "foreignness" could then act in an immunogenic way such that the immune system is stimulated to attack the sun-burned cells. Down-modulation or suppression of this attack would give the skin time to rid itself of these photoantigens, in other words, allow time for skin to repair. Unfortunately, if one of the sun-burned cells is actually a malignant cell, the immune attack normally arising against such tumors would inadvertently be shut down. This is one reason why stressing this mechanism with too much sunlight (e.g. over sunbathing, ozone depletion, UV-B tanning booths) is not to be taken lightly.

When one convolves or multiplies the action spectrum for UV immune suppression with the sunlight spectrum, the resulting product curve indicates we already have levels of UV-B in sunlight necessary for activation of immune suppressor signals under normal ozone levels. This tells us that increased UV-B due to ozone depletion may be expected to increase cell mediated immune suppression. These data also support the idea of the protective nature of this mechanism.

I was asked to give some advice on how to protect ourselves from overstimulating such a mechanism, and the answer basically is just to use common sense when you are out in the sun. It is the same advice we give to people about skin cancer. Avoid the noonday sun plus or minus a couple of hours; wear sunglasses with UV protection and avoid sunglasses that do not absorb UV. If sunglasses are just dark glass this can be more harmful because this will simply dilate your eyes and let more UV through. Wear a hat and wear poor UV transmitting clothes. Be judicious about sunscreen use. From our work and many others, many of the sunscreens are not very effective again immunosuppressive events. For mountain climbers, remember the higher you climb the higher the UV-B; and for swimmers, UV-B can be transmitted down to tens of meters in clear water. In summary, be sun smart and respect the power of the photon.

How can we find out more about UV effects? We have two projects currently on-going, a proposal for Ultraviolet International Research Centers (UVIRC's) and the Scientific Committee on Problems of the Environment (SCOPE) and International Arctic Science Committee (IASC) reports (see References below). Many of you may be familiar with the SCOPE and IASC reports which are implementation plans for research in this area. The Ultraviolet International Research Centers we propose are to be spread around the Arctic and designed to study UV effects on health, aquatic and terrestrial systems and are centered at measuring and monitoring UV-B stations throughout the Arctic.

Finally, many scientists, myself included, have been calling for a focused coordinated UV-B effects programs but because of many factors, not the least of which is a lack of money, we have not really been very successful. Thus it is difficult to provide the answers to the questions people are now beginning to ask in terms of what a given level of ozone depletion, and hence a given amount of UV-B radiation, means in terms of risk assessment. It is important to realize that over eons of time, most living systems have been able to cope with normal fluctuating levels of UV radiation and have successfully adapted to these very energetic rays of sunlight. The concerns now are: To what extent have we already increased the flux of solar UV-B? To what extent will this continue to happen, and for how long? And, how much tolerance do living cells have, or need, to cope with increased levels of UV-B due to ozone depletion, which is almost certain to occur as we move into the next century? The answer at the moment is that we simply do not know. In all candor, I have to say that the effects aspect of ozone depletion is the least funded and not surprisingly, the least understood aspect of this environmental problem. If we are to make any accurate or useful risk assessment on UV-B effects, we must begin to collect essential baseline data in all areas of potential impact before this critical information is lost forever. Basic research on the mechanism of UV-B effects must become a priority since it is knowledge through basic research which will be essential in coping with this problem.

In terms of immune suppression, skin cancer and UV-B, it is conceivable to me that if more emphasis could be put into understanding basic mechanism of these effects, particularly at the level of UV on antigen-processing as I have described here, we may one day be able to switch off those immune signals causing the suppression of tumor destruction. It is entirely conceivable that in this way we might be able to restore the immune system's capacity to attack and destroy skin tumors before they become dangerous or deadly.

References

Special reports on biological effects of stratospheric ozone depletion

De Fabo, E., ed.1997. *Ultraviolet International Research Centers: A proposal for UV-B effects research in the Arctic.* Oslo (Norway): International Arctic Science Committee.

IASC (International Arctic Science Committee). 1995. "Effects of Increased Ultraviolet Radiation in the Arctic." In *Proceedings of a series of IASC workshops.* Copenhagen, Greenland, and Washington, D.C. (1993-1995).

SCOPE (Scientific Committee on Problems of the Environment). 1992. "Effects of Increased Ultraviolet Radiation on Biological Systems." In *Proceedings of a SCOPE Workshop on Effects of Increased Ultraviolet Radiation on Biological Systems.* 17-22 February. Budapest, Hungary.

SCOPE (Scientific Committee on Problems of the Environment. 1993. Effects of Increased Ultraviolet Radiation on Global Ecosystems. In *Proceedings of a SCOPE Workshop on Effects of Increased Ultraviolet Radiation on Ecosystems.* 28 September-3 October. Alghero, Sardinia.

EFFECTS ON SKIN AND EYES

Jan C. van der Leun

It is my task to speak about effects of increasing ultraviolet radiation on the skin and eyes. Why these two organs out of many that we have? The radiation we are talking about, the ultraviolet B, is very strongly absorbed in our tissues, mostly in the upper 1/10th of a millimeter of tissue. That limits the direct effects to the skin and the eyes. In spite of this, it can also have consequences in the entire body through influences on the immune system. Even there, the primary action of the ultraviolet B radiation takes place in the skin.

If you inventory the effects of ultraviolet radiation on skin you find a long list. It includes the obvious effects of sunburn and tanning, but also the formation of vitamin D3 in the skin. There is an aggravating influence on certain diseases, for example, lupus erythematosus, and a curative effect on other skin diseases such as psoriasis. There is also the photodermatoses, a collection of skin diseases where the lesions are caused exclusively by light, primarily sunlight. For most of these, ultraviolet B radiation is an active component. Long-term effects include accelerated ageing of the skin and skin cancers. From the viewpoint of the skin, ultraviolet B radiation is by far the most important component of sunlight; and it is precisely this component that increases with depletion of the ozone layer.

There are also a number of important effects on the eyes: snowblindness, cataracts, and a few less well known effects that are nevertheless serious and can lead to blindness. The most important is cataract. There are indications that ultraviolet B radiation is one among several causative factors. Some time ago, the World Health Organization estimated that seventeen million people in the world are blind because of cataracts. If this increased by only a few percent, it would be a human tragedy. The effects would occur mainly in the most sunny areas of the world and be most severe in the poorest areas with limited medical facilities. That is different from skin cancer where the most serious effects occur in people with light coloured skin who tend to live in areas with better medical facilities.

Our bodies have defense systems against all these effects. Firstly, protection. For the eyes the protection comes from shading the eyes, by squinting, and by contracting the pupil. For skin from hair, from thickening of the outer layer of the skin, and from pigmentation. These are quite effective systems. Then, there is repair. For the genetic material in our cells, DNA, there are at least three systems operating to repair the damage done by this radiation. Finally we have removal. Cells having had too much ultraviolet radiation damage may be removed from the system by a kind of programmed cell death, or by the immune system that recognizes the cells as changed.

This complicated state of affairs makes it difficult to predict whether or not a particular effect will increase with increasing ultraviolet radiation. If we have more radiation, the primary reaction will go up, probably the adaptation reactions also, but the repair systems themselves consist of biological molecules which may also be damaged by the radiation. The immune reactions may be suppressed by ultraviolet radiation so there may be both positive and negative influences. The overall outcome needs to be determined; not all effects of ultraviolet B radiation will increase with increasing ultraviolet B in sunlight.

How do we address this type of problem? One way that initially appeals to most scientists is to do a full investigation of the effect—how it works with all the intermediate steps in quantitative terms—and come to a conclusion by calculating through the entire sequence of events. The problem is that this will take a long time to complete, much more time than we have. Moreover, should we have forgotten one intermediate step, or discover a new reaction, we would have to start all over again. So this is not a practical solution. A simpler way is the "black box" approach of comparing different populations living in areas with different ultraviolet B. The way in which the irradiation leads to the reaction of the skin is not considered in detail. For example, if we have immigrants from northern Europe to the Mediterranean we can study how they react. Doing this, it is evident that they do not suffer from sunburn appreciably more than the people in their home countries, at least, not if they have settled there and are adapted to the new situation. This may be due partly to changes in behaviour, but the adaptation is mainly brought about by thickening of the outer layers of the skin. So, if we had a gradual increase in ultraviolet radiation through thinning of the ozone layer, there is no reason to expect much more of a problem from sunburn. Adaptation could easily deal with the gradual change. Current seasonal changes are much more drastic than the changes that might happen due to ozone depletion.

While this "black box" method is powerful it can also be deceptive. If you compare two areas with different ultraviolet B regimes it is likely there are also other differences. For example, superficially one might be tempted to link the high prevalence of infectious diseases in the tropics, compared to mid-latitudes, to higher ultraviolet radiation levels. But there are many other differences; temperature, humidity, sources of infection, vector organisms and medical care. Amidst such complexity, any influence of the difference in ultraviolet radiation may well be a minor factor. So the method has to be used with great caution.

The final feasible approach is to start with the "black box" method, but take into account all additional information available such as I mentioned for infectious diseases. This is what we are trying to do for all these effects of ultraviolet B radiation.

One general conclusion is that for an effect to increase with increasing ultraviolet B there are three necessary conditions. First, the effect is dominated by ultraviolet B. Second, there is a positive dose-effect relationship. Third, the defenses we have do not take care of the increase. Applying this, we find circumstances where the effects increase with increasing ultraviolet B, some where there is little change such as sunburn as I have already mentioned, and even some where effects decrease.

The latter include most of the photodermatoses, the skin diseases where the complaints are caused by light. These diseases occur most in areas with dark winters; loss of adaptation to sunlight is usually an important component. Many patients suffering from photodermatoses are treated successfully with careful exposures to ultraviolet radiation in winter. This is precisely what would happen with ozone depletion; there would be more ultraviolet B radiation, especially in winter. So these patients would be better off. A similar conclusion applies to the small groups of people with deficiencies of vitamin D.

The effects which increase with increased ultraviolet B are skin cancer, skin aging, snowblindness, and cataracts, although there are uncertainties in the latter. Infectious diseases can increase or decrease as the immune system is very complicated. The influence on the efficacy of vaccinations is undetermined. Overall the picture is still far from clear. Skin cancer is the only effect for which we have good quantitative knowledge. We know a little bit about cataract but it is full of uncertainties. There are also uncertainties with one type of skin cancer—the melanomas—and for the rest we have no quantitative information whatsoever.

There was an attempt to predict consequences of ozone depletion as early as 1970. That was even before chlorofluorocarbons (CFCs) came into the picture. At that time, the supersonic aircraft designed to fly in great numbers within the ozone layer caused concern. An atmospheric physicist, James E. McDonald, singled out skin cancer for calculations. He had very little data but by dredging whatever data were available in the literature, and consulting with dermatologists, he made a quantitative prediction for non-melanoma skin cancer. It is one of the remarkable feats of science around the ozone depletion problem that a scientist, not even a health specialist, could pick out just the one effect where there was any prospect of coming to a conclusion.

Now that we have reasonably good quantitative data about skin cancer, we can use it to make predictions, not because we think skin cancer is the most important effect but because here at least we can do the calculations. Under different assumptions about what would happen to the ozone layer, we can predict through the next century. Considering excess skin cancer cases for northwestern Europe, with no restrictions on ozone-depleting substances the skin cancer rate would be four times its present value by the end of the twenty-first century. With the controls in the original Montreal Protocol, although they reduce the number of cases, there is still a runaway effect and a doubling of the incidence. Only for the later amendments, such as those at Copenhagen, does the curve turn around and the excess disappears by about the end of the twenty-first century. It is remarkable that in the latter case the maximum occurs around the year 2060, much delayed compared to what happens in the ozone layer because of latency in skin cancer. The effect is still significant, and that is on the most optimistic assumption that every country sticks to the rules and that there are no complications and no new threats to the ozone layer. We know already that things are not going according to this optimistic scenario; the more we fall behind, the worse the outcome will be.

A few concluding remarks. The natural sciences first discovered the ozone layer. They also discovered that the layer was vulnerable and came to certain conclusions on what kind of consequences damage to the ozone layer might have. This was all before there was a problem. It resulted from science without an explicit mission for problem solving, just driven by the curiosity of the human mind. This colloquium is in part at least intended to draw lessons from what happened in the ozone layer case, and I would like to transmit one lesson on science and society. There is a tendency in society to ask of scientists and their organizations to devote themselves to solving the problems of humanity. Recently, this was quite explicitly done in the Seoul Declaration for Environmental Ethics. There is good reason for such pleas, but I do not think science should totally comply. It is in the best interests of humanity that there remains a place for science without any constraints. Not directed by a mission, not for solving problems. In the very best case, scientists could indeed solve existing problems, but who would signal the new ones? And that is much more important.

The history of science around the ozone layer demonstrates that without free and curious scientists we would not even know there is an ozone layer, let alone that it could be damaged by our actions. This celebration of the Montreal Protocol's Tenth Anniversary reminds us of the century and longer history of science achievements that were a prelude to 1987; of the success in the past decade of building the knowledge that has dictated stronger action; and of the challenges ahead of sustaining the effort and moving on to realize the benefit.

UV INDEX: A TOOL FOR PUBLIC RESPONSE

Anne O'Toole

INTRODUCTION

Following the discovery of the Antarctic ozone hole and the signing of the Montreal Protocol in 1987, public concern in Canada over ozone depletion and its possible effects began to grow. Media reports contained alarming, and often inconsistent announcements concerning the nature of ozone depletion. Dire predictions on the future of the ozone layer and concerns over the rising rates of skin cancer were being voiced. In Canada, during the late 1980s and early 1990s, public opinion polls consistently showed ozone depletion as the number one issue of environmental concern. There was, however, little information available to the public on the nature of ozone depletion and UV radiation or on what individuals could or should do. There was no information on which to make personal decisions. The search for some way to provide information to the public and to help them to make informed decisions based on expected UV values became a government priority in 1991.

UV INDEX DEVELOPMENT

The Atmospheric Environment Service (AES) of Environment Canada was assigned the task of developing a program which would provide the information needed. In addition to being the home of Canada's weather service with its established operational communications, observing, and training systems, the AES also conducts research into atmospheric ozone and UV-B radiation measurement.

Building on these strengths the AES engaged many other interested parties in the design of the UV Index and in defining the health linkages.

Partners and stakeholders from national health agencies, medical associations, pharmaceutical companies, instrument manufacturers, and the media agreed to help in developing strategies for delivering the product and the public education which accompanied its introduction. Focus groups were also used to test various formats to ensure that the service would be well received by the public.

THE UV INDEX

The result of the cooperative effort was the introduction of the UV Index in the spring of 1992. The UV Index represents the incoming instantaneous flux of UV radiation in milliwatts/m^2 weighted to the International Commission on Illumination (ICI/CIE) erythemal spectrum divided by 25 milliwatts/m^2. UV Index values range from O to 10 in Canada (O–12 in the tropics).

UV Index forecasts are issued twice daily for over forty locations across Canada and are valid for solar noon. The UV Index is classified as *Low*—less than 4; *Moderate*—between 4 and 7; *High*—between 7 and 9; and *Extreme*—greater than 9.

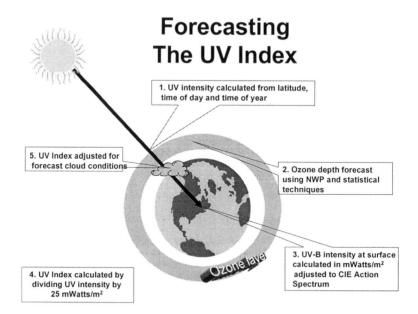

Figure 1. Forecasting the UV Index

The UV Index forecast is prepared for each location by calculating the incoming erythemally weighted horizontal irradiance using forecast total column ozone and

adjusting the result for expected cloudiness (see Figure 1). The total column ozone is calculated as follows:
1) Total ozone is estimated over all of the Northern Hemisphere using a regression relationship between upper atmospheric variables and total ozone;
2) A correction is made to the forecast based on current measurements of ozone over Canada. The procedure runs in a "perfect prog" real time forecast mode with values of the atmospheric variables taken from the Canadian numerical weather prediction model (Burrows et al. 1994);
3) UV Index observations are made at ten locations across Canada using ground-based Brewer Spectrophotometers (Figure 2) and at a number of other locations using various broadband instruments.

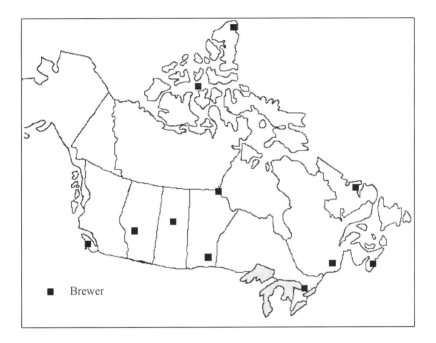

Figure 2. Canadian Surface Ozone Stations Brewer Spectrophotometers

The national UV Index bulletin lists the forecast UV Index for selected cities, indicates whether the Index is expected to be reduced by clouds, and gives the times during the day when the UV Index is expected to be greater than 4 (Table 1). A *Sun Tip* selected from a list of sun smart messages is also included in the UV Index Bulletin as is a chart showing the estimated time it would take type II skin to begin to sunburn.[1] The sun smart messages are derived from the following basic health strategies:

1) Seek shade and avoid the mid-day sun (especially between the hours of 11h00-16h00);
2) Cover up—wear a hat and clothes that cover the arms and legs;
3) Wear UV rated sunglasses;
4) Use a UV protective sunscreen.

Table 1. The daily UV Index Bulletin

FPCN48 CWAO 070745

ENVIRONMENT CANADA FORECAST ULTRAVIOLET (UV) INDICES FOR SELECTED CANADIAN CITIES FOR THURSDAY AUGUST 07 1997. THE UV INDEX BULLETIN IS ISSUED TWICE DAILY EVERY 12 HOURS.

THE FORECAST UV INDEX GENERALLY INDICATES THE UV INTENSITY IN FULL SUNLIGHT AT MIDDAY. ULTRAVIOLET INTENSITIES ARE LOWER UNDER THICK CLOUD COVER AND/OR PRECIPITATION.

VARIABLE CLOUD DAYS ALLOW FULL ULTRAVIOLET EXPOSURE DURING THE SUNNY PERIODS.

LOCATION	WEATHER	FCST UV INDEX	CATEGORY	TIMES UV ABOVE 4
RESOLUTE	FLURRIES	1.8	LOW	NIL
VANCOUVER	MAINLY SUNNY	6.8	MODERATE	10 TO 4
TORONTO	PARTLY CLOUDY	7.5	HIGH	10 TO 4
OTTAWA	ISOLATED SHOWERS	5.5	MODRT DUE CLD/PRECIP	11 TO 3
MONTREAL	CLOUDY PERIODS	6.8	MODERATE	10 TO 4
ST JOHNS NFLD	CLEARING	6.5	MODERATE	10 TO 3

* ESTIMATE OF UV UNDER CLOUD AND/OR PRECIPITATION *

ALL TIMES MENTIONED IN THIS BULLETIN ARE IN LOCAL TIME.

SUN TIP:

BEWARE OF REFLECTED LIGHT FROM SNOW, SAND, CONCRETE OR WATER. BE SMART - USE A SUNSCREEN.

UV CATEGORIES	UV INDEX RANGE	AVERAGE TIME TO BURN
EXTREME	9.0 OR HIGHER	LESS THAN 15 MINUTES
HIGH	7.0 TO 8.9	AROUND 20 MINUTES
MODERATE	4.0 TO 6.9	AROUND 30 MINUTES
LOW	LESS THAN 4.0	ONE HOUR OR MORE

NOTE: AVERAGE TIME TO BURN ONLY ADDRESSES UV EFFECTS ON THE SKIN. UV ALSO AFFECTS THE EYES.

These messages and the order of the advice were determined through broad consultation with the medical community.

The UV Index is disseminated daily to newspapers, radio, television and via the cable TV Weather Network. In most areas of Canada, it is an integral part of the regular weather forecast.

The objectives of the UV Index program are to
1) increase public awareness of variations in UV values;
2) support health agency goals of educating the public on UV risks;
3) assist individuals in adopting a healthy life-style.

While the initial impetus for developing the UV Index was in response to the threat of global ozone depletion, it became clear through interaction with health professionals that over-exposure to UV radiation was itself, inherently unhealthy. Although increased UV levels are associated with stratospheric ozone depletion (Kerr and McElroy 1994), these rises are relatively small and recent.[2] It is more likely that the rising levels of skin cancer being experienced today are the result of changing lifestyles during the past twenty-forty years. However, the increased awareness of UV levels supports the appropriate lifestyle changes required to reduce the risk of skin cancer.

Proper interpretation of the UV Index is important to ensure that people keep the risk factors and responses in perspective. For that reason, the UV Index is trade-marked in Canada to protect it from misuse. It is freely available for adoption by other countries.

Public Reaction

From the outset, public reaction to the introduction of the UV Index has been generally positive. As the public becomes used to hearing the daily UV Index, they are becoming aware of those conditions—time of day, year etc., when UV can be expected to be of concern. This complements other educational efforts by the Canadian Cancer Society, the Canadian Dermatology Association, and other health associations. A public survey taken across the country shortly after the launch showed that 73% of those polled had heard of the UV Index and that of those, 59% were taking some action to avoid over-exposure (Decima 1993)

A more recent poll in 1996 reported that 90% of respondents had seen or heard information about the UV Index. Fifty-seven percent of those polled took extra precautions when the UV Index was high. The same poll showed that 56% believe that sun burning is likely when the UV Index is high (Lovato et al. 1997).

The UV Index is carried by media in most communities and is now mentioned in conjunction with the weather forecast in most areas. This is likely to be significantly raising the public's awareness of the dangers associated with UV radiation.

EVOLUTION OF THE UV INDEX

Time To Burn

The difficulties associated with the *Time To Burn* concept were identified early but it was initially thought to be a useful way for people to relate the UV Index to a familiar effect—i.e. sun burning. However, after the first year, the advice from the medical and health community was to significantly downplay *Time to Burn*. Variations in skin types in Canada rendered this concept inappropriate for those with other than type II skin. There was also a concern that people would use the *Time To Burn* to gauge their maximum safe sun exposure or, worse still, to maximize sun exposure for tanning purposes. Most dermatologists feel that people should be minimizing their sun exposure especially during peak sun periods. Therefore, since 1993, the *Time To Burn* has not been promoted, although it is still carried on the UV Index bulletin.

Appropriate Action

One concern regarding the UV Index program is the difficulty in crafting appropriate messages for different UV Index values, i.e. what does, or should one do differently for an index of 7 versus and index of 9? The basic health messages to avoid the sun, cover up and use sunscreen apply in both cases. Although some countries, notably France and the U.S.A., suggest specific actions for the different UV Index categories, no set of actions acceptable to Canadian stakeholders has been devised. Stakeholders have, however, been strong in supporting the UV Index program as being an excellent vehicle for maintaining public awareness of the need to be sun smart when UV radiation is expected to be significant.

Effect of Clouds

Initially, the UV Index was forecast for clear skies only. This resulted in little variation in the day-to-day forecasts and there was a risk that it would become boring to the news media. This lack of variation was sometimes used as a reason by the media for not carrying the UV Index. As well, a more useful indication of UV levels could only be obtained if the large effects of clouds were included in its determination.

In 1993, a method for adjusting the index for expected cloud conditions was introduced. This system was somewhat coarse and did not completely address the attenuation of UV introduced by clouds. In 1996, an improved cloud attenuation algorithm, utilizing CART (Classification and Regression Tree) statistical methods was introduced which has resulted in more accurate UV Index forecasts (Burrows 1997). This has resulted in more realistic day-to-day variations in UV Index values which are now being carried by the media.

Adoption In Other Countries

The UV Index is relatively easy to apply in other countries. While the index varies from 0 to no more than 10 in Canada, it is open-ended and can be used anywhere including countries closer to the equator where UV radiation is naturally higher. To date, over twenty countries at various latitudes have implemented similar programs using the Canadian UV Index. In 1995, the World Health Organization and the World Meteorological Organization recommended the Global Solar UV Index for adoption by its member countries which is based on the Canadian model (International Commission on non-Ionizing Radiation Protection 1995).

CONCLUSION

Since its inception in 1992, the UV Index has evolved from being a new and unique environmental product to becoming a standard, recognized component of the daily weather forecast in many regions of Canada. It has met its original objectives:
1) It has raised the awareness of Canadians to the dangers of over-exposure to the sun.
2) Health agencies have praised the UV Index as a great help in getting the message of taking sun smart precautions across to the public.
3) It provides information which enables the public to gauge when it is important to take sun safe actions.

The UV Index represents the first attempt in Canada at providing an environmental prediction service. Its success has shown that such products, when scientifically based and developed in consultation with a broad range of stakeholders, can garner public acceptance.

Notes

[1] Skin types are classified from I to VI. Type II always burns easily and tans minimally.
[2] Ozone depletion is generally thought to have begun in about 1980.

References

Burrows, W. R. 1997. "CART Regression Models for Predicting UV Radiation at the Ground in the Presence of Cloud and other Environmental Factors." *J. of Applied Met.* 36: 531-544.

Burrows, W. R., M. Vallée, D. I. Wardle, J. B. Kerr, L. J. Wilson and D. W. Tarasick. 1994. "The Canadian Operational Procedure for Forecasting Total Ozone and UV Radiation." *Met. Apps.* 1 (3): 247-265.

Decima Research. 1993. *An Investigation of Canadian Attitudes Related to Environment Canada's UV Index*. International Commission on non-Ionizing Radiation Protection. 1995. *Global Solar UV Index*. ICNRP-1/95.

Kerr, J. B. and C. T. McElroy. 1994. "Response to Analysing Ultraviolet-B Radiation: Is There a Trend?" *Science*. 264: 1 342-1 343.

Lovato C., J. Shoveller, L. Peters, and J. Rivers. 1997. *National Survey of Sun Exposure and Protective Behaviours. Technical Report.* Vancouver: University of British Columbia (Institute of Health Promotion Research).

WHAT SHOULD BE DONE IN A SCIENCE ASSESSMENT

Daniel Albritton

Let me give you a brief overview of my plan for this summary. The charge to this panel, to myself in particular, was to ask retrospectively the question, "What role has the assessment process, that is the review and taking stock of what we know, played in the Montreal Protocol and how has it shaped it and assisted it?"

The way I will summarize this is to address the role of the assessment panels of the United Nations Environment Programme (UNEP) and of the World Meteorological Organization (WMO) that coordinated the science community in preparing the assessment reports, and look at how they helped shape the evolution of the Protocol and its amendments and adjustments. Most importantly, I will look at what we have learned about how science and research relate to public policy from the Montreal process.

I am going to emphasize the latter point because many of us here and worldwide are involved with other environmental issues. Global warming has been mentioned. It is an issue that will occupy many and even our students and perhaps children over the next decade in managing it. What have we learned from the Montreal Protocol and ozone issue that may guide us as to what it means or what it does not mean for global warming? So you will find that while I am speaking about ozone, I will try to draw some analogies that do work and some that do not work for the global warming process.

Firstly then, you have heard the word assessment. What are they and how do they fit into research and policy-making? The research communities associated with all of our environmental issues have, as their driving force, a better understanding of things like ozone depletion, global warming, biodiversity loss, and impacts of aircraft on the atmosphere. All of these are science phenomena, but they relate to the well-being of humans on the planet. There are those who have to make decisions associated with those issues; decisions ranging from: "Is it a big deal or a little deal?" "How much does it cost if I do something and how much does it cost if I do nothing?"

The assessment process customer base includes governments, which we tend to think of first because decision-making in a public-welfare sense lies there. Industry are clients because they have another crucial role of helping find solutions and in making decisions as to investment portfolios that help address environmental issues. Lastly, as it has been pointed out by several observers, the ultimate decision-maker is the community at large. I am going to touch upon how the assessment process has and has not reached all three levels of that customer base.

So, quite simply, the assessment process is a communication device. As is well illustrated in the case of the Montreal Protocol, it is taking stock of what we know at the moment to help as input to the decision process.

The Protocol has three assessment panels: science, technology and economics, and impacts. Each panel has played a critical role in the Protocol. What I will try to do is to focus on the component of it that I have been involved with in terms of trying to state what we do know and do not know about ozone science for those who want to know it and those who plan to use that information to take some form of prudent action.

Many think of the assessment as producing a report—and it does. This is what I will call the assessment product. If one is asked how well do we understand something at present (stratospheric ozone or climate change), the report will contain a spectrum of sub-statements. Frequently the reports arrange them in this form. From things that the science community is absolutely certain about, so that there are no more papers questioning it, down to things for which, as is often the case, no statement can be made yet because of lack of information. Between these, the assessment reports have included highly confident observations, high-trust calculations, perhaps even just the result of the best tool available, knowing that it is imperfect. They also include scientific judgement calls. I will try to indicate to you examples of all of those because at any point in the development of an environmental issue, there are policy implications of what you are absolutely certain about and there are policy implications of those things about which you cannot yet make a statement. I will try to give you examples of both.

The most recent example from ozone science for the Protocol is the 1994 report done under UNEP and WMO. It was written and prepared by hundreds of ozone research scientists worldwide, either as authors or reviewers. I am very pleased to acknowledge my co-Chairs in this endeavor all three of whom are here today. I invite them to supplement where I may leave some gaps.

Ten years after the Montreal Protocol, can we put our finger on how science shaped it? I say yes, and let me do that in the form of a time line, beginning with the 1974 work from Mario Molina, the discovery that chlorine compounds can deplete ozone very efficiently because of catalytic chemistry. Between that time and when the Protocol was signed in 1987, there was what I would call, an enormous "paleo-ozone" era of patient unraveling of how chlorine interacts with ozone molecules. Indeed, it was research over that period that led to statements such as: "If one continues to add chlorine to the atmosphere, it is very likely that significant ozone depletion lies ahead and the reversibility will be very slow". If one statement had to be representative of the science input to the Protocol, it would be that sentence.

Based on that, the effects on ecosystems and human health, and on the possible options of substitutes for chlorofluorocarbons (CFCs), nations put signatures to the Protocol on the 16th of September 1987. But then a process began, and this is what I want to highlight here. The Protocol asked the expert communities in science impacts and technology and economics to come back periodically and give them their best updated statement of the understanding, because the Parties want to use that information to see whether adjustments or amendments to the Protocol are warranted. How did that interaction work?

The discovery, in 1985, of the ozone hole over Antarctica was a pivotal event for science and for decision-making. In 1989, the science assessment said: "The ozone hole is real, we know what causes it—the accelerated destruction of ozone due to ice particles interacting with CFC decomposition products. So, it is a human-produced phenomena". In London, taking information like that and from other sources, the Montreal Protocol Parties decided to move beyond the CFC freeze and 50% cut and introduced new words. And one crucial new idea was stopping making something, phase it out. This made a dramatic difference, as Dr. van der Leun's presentation shows, in the number of predicted excess skin cancer deaths.

The 1991 assessment reported that not only were there losses in Antarctica of major proportions but there were now incremental ozone losses observed all over the globe. That information fed into the meeting in Copenhagen and the phase-outs were moved to earlier dates.

In 1994, science assessment pointed out among other things that, as can be seen from excellent data reported from the Canadian networks, increases in UV radiation, the next to the last step in the cause-effect chain to human impacts, had been observed. This was reported in Vienna where Parties chose to place caps on other ozone depletors like methyl bromide that were not covered at that time.

So you see that the assessment process is iterative, where the best opinion is given at the moment and the best decisions are made and revisited some years later. It is a crucial part of the structure of the Montreal Protocol and we have just begun a new assessment, to be ready in 1998, to feed into Protocol discussions in 1998 and 1999.

There are a number of questions for that assessment. One that we will definitely and obviously report is: Have we seen the total chorine in the lower atmosphere peak and start going down? Have we seen any resulting changes in ozone and when might they occur? What kind of surprises might lie just over the horizon that we might anticipate a little bit? And finally, the issue of accountability: Watching the ozone hole recover, is it doing what we had predicted?

That is a short sketch of how the science has interacted with the Protocol. Probably as important as the product, is establishing a process. Let me take a little time to go through some aspects of that. As you can see, I am moving into what I think we have learned.

There are four important factors in doing a scientific or state-of-understanding assessment. The first one is the expertise of the authors; the second one is the expertise of the authors; the third one is the expertise of the authors; and the fourth one says there are a few other things. And I would like to elaborate a little bit on those few other things.

First of all, they are done by the expert communities and not by decision-makers. They are done by the experts for the decision-makers. They are based on accessible information. If someone would disagree, they know where they can go and look and see what the judgment was based upon. Then they are reviewed by peers—and I should say reviewed, reviewed and reviewed. Having been a journal editor, I know that these assessments get three to five times more review than a classical peer-reviewed journal paper. They deserve it.

Second, assessments are not policy recommendations and they are not research planning documents. Others have a job description to make policy. An assessment is input to that and may also be helpful in research planning. Assessments are statements of what we know and what we do not know and not designing future research efforts. Perhaps one of their greater values to decision-makers is that it is a one-stop source of integrated information on this phenomena, from the science, to the impacts and the options, done in a quasi-integrated fashion. This has been a model followed for the climate regime. The Intergovernmental Panel on Climate Change has established a similar structure looking at aspects of the climate issue.

Third, they need to work for the client or customer. If I had to give grades to our efforts over the last ten years, I would give a grade for the beginning period and I would give a grade for now. We probably have done best interacting with governments. They were the ones to ask the questions first and we began interacting on the policy decision process. In the early periods, scientific researchers had a hard time understanding the business sector role. There was an unfortunate polarization period—which worked in both directions—which has now disappeared, and there is now very good cooperation. That is the lesson that I hope is learned on the climate issue: the shorter the adversarial period, the more fruitful the dialogue. But I think we flunked interacting with the public ten years ago. We are probably doing a little better now. Probably one of the hardest things for the scientists to do is to deal with that ultimate policy-maker using that ultimate policy-maker's vocabulary and terms.

Finally, we learned that science and policy are interactive. And that, I think, is a valuable lesson, namely: there is no final answer in science. There is no final action in policy. There is an improved answer and an improved set of decisions. Understanding this takes a lot of the heat out of the difficulty of taking actions early. We also learned that in talking to policy-makers, rather than presenting fifty choices, present four choices, spread wide enough that there is no doubt that the science is robustly saying that one is different from the other. That too is a lesson for the climate issue.

Now, I would like to switch from the specific lessons we learned about science assessment to a few more general lessons. Again, these are personal views.

I would maintain that there are three phases to the evolution of an environmental issue that are well illustrated by the Montreal Protocol. There is an early phase where the chief job of the science community is to describe what level of credibility is associated with that issue. The second phase is associated with the manageability issue. The best job that the technical and scientific experts can focus on after decisions are made is how one can implement them without breaking the bank or without making a mistake. Finally a third phase is the accountability issue: what does not come up often in scientific planning is planning to be accountable for what we describe in terms of our understanding.

For the ozone issue we are well beyond phase one, the credibility issue. I would argue we moved through that somewhere between 1985 and 1989, and away from questions like whether chlorine destroys ozone, whether the Antarctic ozone hole is natural or human produced, or whether there is a global downward trend in ozone. These issues are settled. They were part of credibility building.

What is the parallel with global warming, which is another instance of atmospheric response to a human-produced molecule? In 1995, the Intergovernmental Panel on Climate Change (IPCC) reports stated that the balance of evidence suggested that the current run up of temperature is probably not all natural. A first tentative statement that was very similar to the one twelve years ago that said human produced chlorine very likely is responsible for the loss in… etc., etc. In fact, many times one can close one's eyes and switch the words from ozone to climate and see an exact parallel. There are some times when it is the exact opposite parallel and I will try to cover that too.

Phase two deals with the manageability issue and how assessments deal with that and what it means to researchers. Obviously, governments can make a decision and say: "Ah! No more CFCs". But somebody has to have an alternate and a product available to fit in the refrigerators, clean the electronics, etc. The whole scientific issue of the ozone friendliness of those substitutes is part of the researcher's role during the manageability phase. The question of whether and how much one gains from a phase-out of CFC substitutes, and of the costs involved, are all manageability issues in which the science community and the technical communities have a role, but a very different one in my opinion from their role during the credibility phase.

Phase three is accountability. This is something I think we are learning slowly how to do. We need to do it more consistently. As part of the accountability, we have already shown that the growth rate of chlorine- and bromine-containing compounds in the lower atmosphere has peaked and has started going down. That is a very good first step in the accountability era of the ozone phenomenon. But, how well can we describe when we should see ozone itself improve, when we should see a weakening of the Antarctic ozone hole? We must remain in position to be able to be accountable for those who believed us back in the credibility phase. That is how the assessment process under the Montreal Protocol is moving with emphasis on those three phases.

The second point I want to leave with you is another general lesson that I think we have learned. That is, that the single most important property of a molecule that we put in the atmosphere is its time constant or atmospheric lifetime. Of course, the molecules have to interact somehow, either react with ozone or interfere with infrared escaping. But, probably the most policy relevant property in the chemical reference handbook is how long that molecule resides in the atmosphere.

Let me give you some examples of this. There is a set of compounds called perfluorocarbons where all of the chlorine and bromine one would find in a CFC or halon are gone and is replaced by fluorine. They are totally fluorinated. What has been found about those compounds is that they are probably immortal. The only way they are removed is probably destruction at the very upper part of the atmosphere by the most violent solar radiation, and perhaps in high-temperature furnaces. But, other than that, when they are in the atmosphere they are there to stay. Their lifetime estimates are fifteen to eighteen thousand years, about the time our species has been here. So without our society yet having manufactured a lot of them, and without yet analyzing all of their consequences, just the fact that we cannot get back to the original condition for fifteen thousand years has led many to say that these are probably not good substitutes for halons.

So, that property and that property alone leads to a very serious decision. Compounds like CFCs and CO_2 have lifetimes longer than our own. We have willed our CFCs to our children and we willed our CO_2 to our children. Releasing chemicals with long lifetimes has bought us ownership of an Antarctic ozone hole for the rest of our lives—maybe not the rest of all our lives; some here will see it disappear. Closing the ozone hole should occur about the year 2050 all other things being equal.

Let me indicate a few other properties about lifetimes. The planet has lifetimes that it chooses to move in. Many are very fast like tomorrow's weather, but ocean circulation moves really slowly. The burying of surface water in the deep ocean, which is where heat and CO_2 are absorbed, has a lifetime of 100 to 500 years. So here is a planetary time constant. If we choose to move the planet along in that direction, it is going to be very slow to move, but once moved it is going to be very slow to come back. And so, there is a time constant in environmental issues that plays a very big factor, just like the CFC time constant does in the Montreal Protocol.

There are also technological time constants, and I raise this example because it is an environmental issue that is emerging, and the assessment process is just beginning. But you know, airplanes can live forever. Almost all of the fleet that was built is still flying with new more efficient engines and better maintenance. I am told by the experts it is only a matter of how much you want to pay to keep them alive. Technically, they can go on and on. The issue here is that if there are decisions associated with aviation emissions, it is not the chemical that comes out of the engines that one worries about. That has a very short timetable. One wants to ask about decisions now regarding the shape of the fleet for the next fifty years, because many of them will be around that long.

The point I want to make on lifetimes is the following. A decision-maker has to decide when there is enough information to risk making a decision which has some upside and some downside. I would argue that for phenomena that have very long time consequences, that balance in that equation is different than for phenomena in which there are very short time constants. The long time constant says you have discovered you are in a poker game and the deck is rigged and you cannot quit for a very long time. A short time constant says I may wait for more information because I know that when I do make a decision the phenomenon will reverse itself fairly rapidly. These are two very different classes of environmental issues.

Finally, let me conclude by pointing out some explicit similarities and dissimilarities between the ozone depletion issue and its Montreal Protocol, and the global climate change issue and the discussion about a protocol that will occur in Kyoto. There are many things that are one-to-one analogies, there are other things that are one to minus one analogies. And it would be important in this ten-year anniversary of the Protocol to have a little idea of where it could help as an analogy and where it could probably hurt as an analogy. Let me go through at least a few opinions on that.

There is a difference between local phenomena and global phenomena. A city may decide on its river quality or on its reservoir quality. A nation or a group of nations may decide on transboundary issues, but on a global issue like ozone depletion, what we have noted from the Protocol is that there has to be a global buy-in. That is a similarity between greenhouse warming and the ozone depletion.

I already mentioned the time constant. I will not belabour that point, but let me indicate some things that are not quite clear yet. There were surprises in the ozone issue. The most dramatic surprise is that we were clueless about Antarctica. The reason was that there was a process lying down there, waiting for chlorine to get up a certain level which it never had done before, and hence we never could have observed it. Once it reached that level, a quadratic dependence got tickled and kicked into the process, activating processes that we had no clue the planet knew how to do. The issue in climate may come down to similar things. Those here who know more about it than I would likely agree that the climate system is probably as non-linear as the ozone stratospheric system. Therefore, thresholds and surprises, both rude and pleasant, lie ahead in the climate issue just as the biggie came along in stratospheric ozone. So, when the question comes up about surprises, I think the answer is likely yes, even though one cannot define of course what you do not know.

There are dissimilarities where one could potentially make mistakes by relating one to the other. A key factor in the Protocol was that as the science was gelling and the credibility phase was beginning to get solid, and questions of where to go and what to do were pressing on policy-makers' minds, industry developed and brought forward relatively inexpensive substitutes for the compounds that were beginning to wear a blacker and blacker hat. The availability of a way out due to the ingenuity and the cleverness of industry was a key factor in that balance I tried to sketch for you between the cost of doing something and the cost of not doing something. It is not clear what are the HFCs and hydrochlorofluorocarbons (HCFCs) for carbon. One can name them but one can also name huge drawbacks or advantages depending on one's political point of view. The discussions of where to go if decisions are taken on the greenhouse issue will be much more complex than they were on stratospheric ozone.

And mainly, the difference in industrial economic scales of the two issues are staggering. To be sure, important sectors of the economy were influenced by the CFC decisions. Important decisions were made. They influenced the course of major companies. In the climate issue, the decisions of not doing anything have huge costs, maybe a thousand-fold more than those associated with ozone. The decisions of actually doing something also have that scale because of the fact that the decisions are very expensive and "no decisions" could well be equally expensive. The key point I want to make about assessments, and then I will conclude, is that the more complex the issue, the more demands it is putting on a clear understanding, and the value of straight-shooting, unbiased information goes up and up and up in the decision process. And that, I think, is the key reason why all issues need some form of assessment process.

Let me stop here with one little set of personal remarks. This is a series of three ways of saying what I think we have learned about the science and the policy interface. I thought I discovered some of them and I later found out from my friend Dick Sullen, who gave me a copy of a little book written by Walter Lippmannn, a newsman and commentator, now dead unfortunately. In that book, *Public Opinion*, he wrote an article on the role of the expert in decision-making. It is a very impressive and penetrating essay, pertinent here even though it has nothing to do with science. I will put a science spin on it.

First of all, science does not make public policy. The squirmiest you see a scientist is when you ask them what we should do. Because if they have thought about it, what you do depends not only on the science, but technological options, the economics of a versus b, the politics of sovereign nations trying to decide on one global phenomenon, and considerable diplomacy as we have seen. What Lippmannn recognized back in the 1920s was that it is no accident that the best service in the world is the one in which the divorce between the assembling of knowledge and the control of policy is the most perfect.

He also then goes on to point out that while the information producer should not make the decisions, and the decision-makers should not be the information producers themselves, those who investigate and those who decide must have lunch together. And the assessment process is a way of having lunch together. That is, the research community, or a small fraction of it, must interact with the decision-making process. I would argue that the Montreal Protocol has set an example of that and the Framework Convention on Climate Change is trying to establish that.

Finally, what it means to us as researchers. I hope there are researchers who are asking how can I do my research and also feel that I have made a difference, because I have an answer for you. Scientific insights are a service. What Lippmann said was the idea that an expert is an ineffectual person because he or she lets others make the decision is totally contrary to experience. The Montreal Protocol shows he is right. Namely, that maximum chlorine in the lower part of the atmosphere peaked at 3.2 ppb in late 1994 or early 1995 is attributable to both the experts and the decision-makers. *That* is a service, and *that* is being effective.

COMMENTARY ON PRESENTATION BY DANIEL ALBRITTON

Gordon McBean

First of all, I would like to say that it is very hard to follow Dan's excellent presentation because he is not only an authority in the field of science assessment, he is also very good at explaining them. I think that it is very clear that the role of the scientific assessments have been very important in the ozone issue and the climate issue. The reasons have been already enumerated by Dan, but I would just like to add to or reinforce some of the things he mentioned.

On the whole question of international credibility, as he said, it is the expertise of the scientists that is the basis of that credibility. It has been very important in the assessment processes, both for ozone and now for climate change, that these assessments have been reviewed and re-reviewed by a very broad range of scientists in all parts of the world, from countries with highly developed scientific communities to ones where it is, unfortunately, less so. International development of consensus has been the basis that has allowed international conventions to move ahead. If there had not been, or if there were only a small number of scientists involved, the support for the development of these international conventions and their amendments would be much weaker.

The other thing that has been very important is that the various assessments have gone to the next step beyond where we scientists have usually stopped. They have developed something called a summary for policy-makers which tried to convert from the usual scientific jargon that we tend to write, to a wording, a presentation that is expressed in a way which policy-makers, or at least the non-scientists, will understand.

Although I have not been involved in the scientific assessment process for ozone, I have been very much involved in both the 1990 and 1995 assessments of the Intergovernmental Panel on Climate Change. The 1995 scientific assessment and summary for policy-makers—that very critical document which was originally supposed to be twenty pages, and I think ended up more like thirty-some—was a highly debated preparation process. We worked that through in Rome where we started on late Tuesday afternoon having already done ninety percent of what we thought was the agenda by noon. I thought we were all going to be able to site-see through Rome at least by Thursday morning. Well, we started going line-by-line through the summary for policy-makers on Tuesday afternoon and we ended at one minute to midnight Friday. We only ended then because the government of the city of Rome said that that building had to be evacuated at midnight and they actually had the police waiting to move us out if we were not out. That was certainly a motivation but I think the importance of these summaries for policy-makers should not be in any way underestimated. It has become very critical in terms of being able to explain to politicians and other so-called policy-makers or decision-makers what is the state of the science.

Dan talked about the comparability of the climate and ozone issues. I agree with all of his assessments. It was interesting that the ozone assessment process seemed to have started after the Montreal Protocol. While there were assessments before, the formal assessments came after the Protocol was signed. Maybe that was one of his minus analogies.

However, it is worth stressing that we should never think of these as truly isolated or separate issues. Climatic change and ozone depletion are very intimately linked. Chlorofluorocarbons (CFCs) are not only ozone-depleting substances but also very potent greenhouse gases. It was interesting that a good piece of the evidence that allowed the climate change community to deduce that there was discernible human influence on climate in the 1995 assessment was bringing in the ozone chemistry and the understanding of how the stratospheric layer of the atmosphere had changed due to ozone depletion. That piece of evidence allowed the finger print method that Ben Santer and others used, to really add to the credibility of the climate statement. So, scientifically, there have been very strong linkages.

Robert Worrest mentioned that increased UV radiation will probably impact, if it has not done so already, on the prevalence of phytoplankton in the global ocean which are a critical part of the global carbon cycle. The interactions between all our environmental issues is something we should never forget.

Let me stress a few other points. It is very important that we recognize the essential nature of basic scientific research. It was mentioned that had we not had a strong basis of science, we would not have been able to understand what was going on in either the ozone case or the climate changes issue. For ozone we had to have that strong basis of understanding of basic chemistry. I think Professor Molina referred to the work of Crutzen on various kinds of ozone reactions which were necessary to understand catalytic stratospheric ozone depletion. It was also important to understand atmospheric dynamics well enough to decide in the end that it was primarily the CFC chemistry role, as opposed to some misunderstood part of the dynamics, allowing one to focus in on the issues in the ozone depletion case. So, all kinds of scientific disciplines come into these issues.

This emphasizes the increasing necessity for the scientific community, across a whole variety of disciplines, to work together to understand and provide advice to policy-makers through science assessments. I was brought up as a scientist, I studied physics, and even though I became an oceanography professor I generally tried to avoid having to know anything about the biology. But one soon discovered that was not possible. There is an intimate link between biological, chemical, and physical processes. Although it is essential that our scientific community be able to focus, so we can move ahead in our niche area, we must also be able to work across disciplines in addition.

A couple of other quick points. I would like to stress the importance of ongoing monitoring programs. In 1957, for entirely pure-science reasons of interest to a community of scientists, the International Geophysical Year (IGY) was held. Most people probably never heard of IGY, but in 1957 it stimulated a series of systematic observations. The Keeling observations of carbon dioxide at Mauna Loa, the British Antarctic Survey ozone measurements, and Canadian ozone layer observation, which had been done sporadically, were all started on a systematic basis. That meant that when ozone depletions were observed in the Antarctic, and later in other parts of the globe, we had approximately thirty years of data. We also have these data for CO_2. There is a tendency in governments to say: "You don't need to monitor the atmosphere anymore, you have already ten years worth of data. Why don't you turn it off?" Well we cannot afford to do that. Unfortunately, we cannot afford to do all the monitoring that we need to do either. So we have to have the right balance, work together, and also very importantly, exchange data among countries in a free and open way.

Lastly, let me comment from my own personal experience. One of the realities is that in order to catch political and public attention, we have to simplify problems. In the middle 1970s, I was chairman of the Environment Canada Scientific Committee in the Long Range Transport of Atmospheric Pollutants. Now, that attracted absolutely zero attention from the public and not much more from the Minister of the time. But when it was decided to call it *acid rain,* the interest of people shot up. It was the same issue but we got a fancy title. I am not sure if someone thought consciously when they referred to the *ozone hole,* but the reality is you get attention when you get a dramatic phrase like that. Global warming in Canada has not yet become a real public concern because we freeze half of the year in this country. Global warming has not yet taken on the urgency of acid rain or an ozone hole. That is a reality we are still living with.

Part 3

THE ROLE OF DIPLOMACY

THE MONTREAL PROTOCOL AS A NEW APPROACH TO DIPLOMACY

Richard Elliot Benedick

It is very fitting that we meet here in Montreal on the tenth anniversary of this landmark international agreement. I would like to begin by paying personal tribute to Canada for its steadfast leadership and commitment for over two decades to the crucial, and often daunting, challenge of protecting the ozone layer.

It hardly seems that ten years have passed since I arrived in Montreal, following the hardest bureaucratic battle of my career: thanks to the unwavering support of Secretary of State George Shultz, we had turned back a last-ditch effort by anti-environmental ideologues within the Reagan administration to oppose a strong treaty—and also to fire me as chief U.S. negotiator—a decision that had to go to the President himself.

I have so many memories of those historic days in Montreal—one of the most vivid is how the distinguished Austrian ambassador and scholar Winfried Lang and I together designed the critical article 2.6 on the back of a menu during lunch, which enabled the then-Soviet Union to join the treaty.

But since this is an academic colloquium, I will not regale you with diplomatic anecdotes. For a full story of the negotiations and analysis of the protocol's history, I might refer you to my *Ozone Diplomacy*, which Harvard University Press will bring out in a substantially enlarged new edition in early 1998.

Instead, today I would like to focus on what makes the Montreal Protocol so special in the annals of diplomacy and international environmental cooperation. The last decade has witnessed a virtual explosion of large multilateral negotiations on the environment. These include:
1) The 1989 Basel Convention on the Control of Transboundary Movements of Hazardous Wastes;
2) the 1991 establishment of the Global Environment Facility;

3) the 1992 U.N. Conference on Environment and Development, and its related Agenda 21;
4) the 1992 Framework Convention on Climate Change;
5) the 1992 Convention on Biological Diversity;
6) the 1994 U.N. Conference on Sustainable Development of Small Island States;
7) the 1994 Convention to Combat Desertification;
8) the 1994 International Conference on Population and Development; and
9) the 1995 U.N Conference on Straddling and Highly Migratory Fish Stocks;

as well as negotiations of the U.N. Commission on Sustainable Development and its associated bodies, and numerous intergovernmental working groups on such subjects as forest management, land and water resources, economic instruments for environmental management, and biotechnology.

These were not unique events, but rather constitute ongoing institutional arrangements and negotiations to appraise the effectiveness of existing international commitments as well as to address new and changing threats to the environment. Taken together, they represent a still-evolving system of international environmental governance—which did not exist at all a short ten years ago.

The Montreal Protocol was in many ways a model for all this subsequent activity, a treaty and a process that incorporated many unique features and whose influence is perceptible far beyond the sphere of ozone protection. I would like to examine seven major precedents and aspects of the Montreal Protocol that made it so special as an international agreement:

1) innovative elements of the negotiating process;
2) the unconventional emphasis on science, even when scientific certainty remained elusive;
3) the dynamic and flexible character of the treaty's commitments, provisions, and interpretation by the state-Parties;
4) the institutional framework that developed after the treaty was signed;
5) the treaty's reliance on market instruments that unleashed a great surge of research and development of new technologies;
6) the sensitivity to equity among sovereign nations, as manifested in new forms of international cooperation and assistance; and
7) the stimulus that the Protocol gave to the creation of a worldwide network of non-governmental organizations (NGOs).

First, the unconventional process: it is not well-known that the Montreal Protocol actually originated from a failure. In March 1985, the Vienna Convention for the Protection of the Ozone Layer was finally agreed after three years of hard negotiations, but it did nothing to control emissions of ozone-destroying substances—it did not even mention chlorofluorocarbons (CFCs)! In the closing hours of the Vienna conference, however, the Nordic countries, Canada, the U.S., and a handful of allies introduced a last-minute resolution that would authorize the United Nations Environment Programme (UNEP) to reopen negotiations with a 1987 target for arriving at a legally binding control protocol. The resolution further provided for informal fact-finding workshops to precede the formal diplomatic negotiations. This diplomatic maneuver appeared to be an afterthought, but it was actually carefully planned. It caught the opponents of international controls by surprise and effectively isolated them. The resolution passed, and thus launched the process leading to Montreal.

A key element of this process was the idea of informal workshops where negotiators could mingle with scientists and representatives from industry and NGOs in a relaxed atmosphere, speaking in their individual capacities and not as government representatives. Meetings such as the one I chaired in Leesburg, Virginia, three months before the opening of formal diplomatic negotiations, provided a unique opportunity to exchange ideas, break down problems into smaller components, explore options, and attempt to narrow the ranges of official disagreements.

The principle of informal workshops and consultations also continued after the signing of the Protocol, helping to build personal relationships among the negotiators and generating creative ideas that facilitated the formal negotiations on the further strengthening and implementation of the treaty. In fact, after I retired as an official negotiator, I continued to apply this technique in organizing and chairing "policy dialogues" to assist the climate change and other environmental negotiations. For example, I have been asked to chair a similar meeting during the First Conference of Parties to the U.N. Convention to Combat Desertification in October 1997.

The Montreal Protocol experience also exemplified a new fact of life for the modern diplomat: he or she cannot, as in traditional diplomacy, negotiate behind closed doors, aloof from society at large. Rather, the environmental diplomat must work closely with industry, with advocacy groups, and with scientists. Indeed, before the Montreal Protocol was completed, my colleagues and I had managed to attain a degree of fluency in atmospheric chemistry that certainly would have astonished my high school science teachers.

This leads to the second theme, the central role of scientists in the new environmental diplomacy. The Montreal Protocol was negotiated at the frontiers of modern science, relying on arcane computer models simulating intricate chemical and physical processes for decades into the future, and on satellite monitoring of remote gases measured in parts per trillion! Without modern science and technology, we would have remained unaware of what was happening fifty kilometers above the Earth, and the results would have been truly catastrophic. Research on the ozone layer also revealed previously unrealized linkages among different scientific disciplines. The treaty and its subsequent implementation were a truly interdisciplinary effort involving chemists, physicists, oceanographers, biologists, soil chemists, toxicologists, meteorologists, botanists, entomologists, engineers, oncologists, and many more.

The development of a commonly accepted body of data and analyses and the narrowing of ranges of uncertainty were the prerequisites to a political solution among negotiating governments that were initially far apart. Scientists were drawn out of their laboratories and into the negotiating process, and had to assume an unaccustomed responsibility for the policy implications of their findings. For their part, political and economic decision-makers needed to understand what the scientists were saying, to fund the necessary research, and to act before all the evidence was in.

The inevitable scientific uncertainty conditioned the third aspect that I would like to highlight for you. Unlike traditional treaty-making which seeks to cement the status quo, the Montreal Protocol was deliberately designed to be modified: it is a dynamic and flexible instrument. The treaty contained critical procedures for reopening key commitments as conditions evolved. Such changes could be made by amendment, by adjustment, or by decisions of the Parties to the Protocol. Once a chemical was placed on the control list, subsequent tightening of controls would not require the lengthy ratification process implied by a formal amendment, but rather could enter into force as an "adjustment". This was a conscious effort to make the treaty responsive to fast-moving scientific revelations of dangers to the ozone layer.

Any changes or adjustments to the Protocol would flow from periodic formal reassessments of the state of the science, the technology and economics of substitutes, and data on environmental effects of ozone depletion. Such reassessments were a formal requirement and a central and unique element of the treaty. In effect, the Montreal Protocol was not a static solution, but rather an ongoing process.

A guiding principle of the Protocol as it evolved over the past ten years was to send the right signal—and to avoid sending wrong signals—to industry as well as to the contracting Parties themselves. Hence the Meeting of Parties and its associated institutions interpreted the treaty flexibly while not doing damage to the fabric of the commitments; this was clear from decisions taken over the years on data reporting, Multilateral Fund issues, nominations for exemptions for essential uses, and the unanticipated problems that arose in the formerly centrally planned economies of eastern Europe. There emerged a concept of the "spirit of the Protocol" which, while never precisely defined, was implicitly understood by the contracting governments as representing their underlying commitment to protect the ozone layer. In this sense, the phrase was actually employed in legal decisions of the Parties.

The negotiators, ten years ago, also did not attempt to resolve every possible contingency. Our highest priority was to get emissions reductions into international law as fast as possible. Because we did not want to delay this process by extended legal wrangling, we deliberately left several unanswered questions at Montreal in 1987, postponing decisions on such important issues as the financial mechanism, non-compliance procedures, and details of the trade restrictions. Incidentally, the trade restrictions themselves were an innovation in an environmental treaty. By ensuring that non-parties could not profit from trade in ozone-depleting substances, the trade provisions proved very effective in encouraging virtual universal membership in the Protocol.

As Mostafa Tolba, then Executive Director of the United Nations Environment Programme (UNEP), wearily declared in his closing speech to the delegates here in Montreal on September 16, 1987, "This Protocol is a point of departure... the beginning of the real work to come." The implementation of this work is my fourth theme: as the most mature of the recent environmental treaties, the Montreal

Protocol developed a sophisticated institutional framework to ensure that the treaty functioned effectively after the ink dried on the signatures.

At the top of this structure is the Meeting of the Parties (MOP), which combines executive, legislative and judicial functions in a single supreme decision-making body. The more than 160 governments that are Parties to the Protocol meet annually to decide upon a wide range of issues ranging from strengthening controls to cases of non-compliance.

Supporting the MOP is the open-ended Working Group, a less formal negotiating body that meets two to three times each year to examine the issues, narrow the alternatives, and recommend decisions for the MOP; many of the same national delegates are involved in both bodies.

The Implementation Committee also reports to the MOP; it comprises ten members from the state-Parties and plays a crucial role in monitoring compliance and in helping those Parties that experience difficulties in meeting their obligations to find realistic and effective solutions to their problems.

The Ozone Secretariat, located at UNEP headquarters in Nairobi, is responsible for documentation, coordination, and logistics for the MOP and most of the other subsidiary bodies of the Protocol.

The critical function of periodic assessments is performed by three Assessment Panels that meet on a more or less continual basis to provide the scientific, technological, economic, and environmental affects assessments that have proven so crucial in the evolution of the original protocol. The assessment process has involved hundreds of experts of dozens of nationalities, coming from universities, governments, research institutes, NGOs and international organizations—comprising an unparalleled body of expertise available to the Parties.

The Technology and Economic Assessment Panel (TEAP) itself has several "technical options committees" that correspond to the various industrial subsectors affected by the treaty (refrigeration, foams, solvents, etc.). These committees evaluate progress and examine technological advances in light of the present and prospective controls over the ninety-five different chemicals covered by the Protocol. Ad hoc groups were occasionally established to address such issues as problems of Parties in central and eastern Europe in meeting their treaty obligations after the collapse of state-planned economies.

The Multilateral Fund for the Implementation of the Montreal Protocol (MLF) has given rise to a related complex of institutional mechanisms. The Executive Committee, comprising fourteen members divided between North and South, is a powerful body that represents the Parties in managing all aspects of the Fund, from approving multi-year budgets to approving projects. The MLF secretariat, located in Montreal, also plays a major role in developing recommendations for the Executive Committee and in relations with the Fund's implementing agencies.

The MLF has four implementing agencies—World Bank, UNEP, United Nations Development Programme (UNDP), and United Nations Industrial Development Organization (UNIDO). Each of these agencies also had, or created, its own institutions or committees to interact with those of the Protocol.

It can thus be appreciated that the actual negotiation of a modern environmental treaty is not an end in itself. Rather, the entry into force of the international accord becomes the springboard for further negotiations and implementation issues, requiring a structure of new institutional arrangements to deal with complex, evolving situations. In the case of the Montreal Protocol, I am grateful for Professor Edward Parson's astute observation that the delicate interplay between the Meeting of Parties, representing governments as the overall decision-making body, and the numerous panels of scientific and technical experts from the outside world, has proven to be a central factor in the treaty's further evolution and success.

A fifth innovative aspect of the Montreal Protocol was the way it encouraged use of market instruments that generated a wave of technological innovation. It is fair to say that no one present in this city a scant ten years ago dared hope this could be possible. Unlike traditional environmental regulations that attempted to compel industry to adhere to "best available technologies", the Montreal Protocol represented a true leap of faith in the market system. For the negotiators in 1987 mandated firm targets and timetables to reduce consumption and production of the offending chemicals, with full knowledge that alternatives were not yet available! And this tradition has been maintained even as new chemicals were added to the list of controlled substances.

The Montreal Protocol was truly a "technology-forcing" event. During the 1970s and much of the 1980s, industry had argued that alternatives to CFCs and halons—"ideal" chemicals used in thousands of products and processes and still finding ever new applications—would be ineffective and enormously expensive, if they could be found at all. DuPont, the world's largest producer, had even shut down a modest research program in 1981, when President Reagan's new administrator of the U.S. Environmental Protection Agency (Anne Gorsuch) proclaimed the ozone layer a "non-problem".

The Montreal Protocol changed all this with the stroke of a pen. It gave a clear signal to industry that research and development in alternatives to the ozone-depleting chemicals would now be profitable, indeed was now essential. In this way, the Protocol mobilized the forces of the market and unleashed the creative energies of the private sector. Instead of obstructing controls, entrepreneurs now found new solutions, often where least expected.

Interestingly, the treaty did not dictate specific actions, but instead allowed individual nations to develop their own systems according to their own specific circumstances. This flexibility led to a wide range of experimentation by the state Parties—including taxes, quotas, product or process standards, product labeling, voluntary agreements, tradeable permits, import restrictions, and other instruments. Policies generally allowed considerable latitude to the private sector to decide on the most cost-effective ways of meeting the reduction targets—a factor that was critical to the success of the Protocol.

A rather unique spirit was born. Governments, industries, and environmental organizations cooperated with each other, sharing the results of their experience. Consortia were established among otherwise competing companies that saved both time and money in research and testing of new alternatives to ozone-depleting substances. Industries that used the critically important CFC-113—which had been considered an indispensable solvent for telecommunications, computers, aerospace, and similar products—did not even wait for the chemical companies, but reexamined their own manufacturing processes and came up with an amazing variety of answers, including new "no-clean" techniques that obviated the need for chemical solvents.[1]

The sheer volume of ingenious new alternatives across the entire range of thousands of products and processes served by the ubiquitous CFCs and their ozone-destroying relatives was astonishing. The Montreal Protocol had triggered a virtual technological revolution within an amazingly short period of time. What had previously been considered impossible became a reality: the "ideal" chemicals were being replaced, and the innovative new technologies and substitutes often proved both better and cheaper. Significantly, initial estimates of the costs of transition away from the ozone-depleting chemicals proved to have been vastly exaggerated.

A sixth unique aspect of the Montreal Protocol was its sensitivity to issues of economic and structural inequities among sovereign nations. We now take for granted the concept of shared, but differentiated, responsibilities among nations of the North and South when it comes to protecting the environment. It is clear that the tremendous and still rapidly growing populations of the South, with their legitimate aspirations to attain higher standards of living, could simply overwhelm international efforts to protect the global environment, whether it is the ozone layer, the climate system, the biological richness of species, or the tropical forests.

The designers of the Protocol recognized that developing countries would, starting from a very low base, need for a certain period of time to expand their use of ozone-depleting substances, but that financial and technological assistance would have to be provided to limit this expansion and promote a phase-out. With the ozone issue, the richer countries of the North for the first time recognized their responsibility to help developing nations to implement the policies needed to protect the global environment. It is not widely known that, for many developing nations, the Montreal Protocol provided the first intensive exposure to environmental problems, leading both to greater awareness and to the development of new institutional capacity.

Elements of the Multilateral Fund became models for the 1991 Global Environment Facility and for the subsequent climate change and biological diversity conventions. The Protocol evolved a uniquely balanced voting procedure for making decisions if consensus proved impossible to attain: separate majorities of North and South that together would build a two-thirds majority. The Fund also pioneered the concept of financing only the "agreed incremental costs" to a developing country of meeting its treaty commitments, and the annals of the debates of the Fund's executive committee clearly reveal that this distinction was fairly applied in many complex situations.

The Montreal Protocol was also the first global experiment in environmental technology transfer, served by a network of industry consortia and informal connections involving governments, industries, and even NGOs, and reinforced by the extensive information clearing-house activities of UNEP's hardworking Paris office. The Protocol's history provides numerous examples of unusual North-South cooperation to promote transfer to the South of the new technologies needed to phase out CFCs. Greenpeace collaborated with a former East German manufacturer to develop an ozone-friendly refrigerator, which was subsequently promoted in China and India by the Swiss and German official aid programs. More than forty companies from eight countries joined to help Vietnam's phase-out. Japanese and American importers of electronic components cooperated with the U.S. Environmental Protection Agency and Japan's Ministry of Trade and Industry to provide technologies to accelerate Thailand's phase-out. Canada's Northern Telecom (now Nortel) played a similar role in Mexico and many other countries. And these are only a few examples.

The Protocol's sensitivity to equity issues is also evident in its treatment of non-compliance. The non-compliance procedure, involving the Implementation Committee, the Fund's Executive Committee, and the Meeting of Parties was, like so much of the Montreal Protocol, innovative, pragmatic and flexible. Rather than punishment, it emphasized cooperation, encouragement, and assistance to Parties in overcoming problems of non-compliance. It motivated the contracting Parties by building on their own good intentions and sense of responsibility. This flexible procedure, reflecting the approach of diplomats rather than Green Berets, proved its value when several countries of central and eastern Europe—including Russia, a major CFC producer—found themselves unable to comply with the 1995 CFC phase-out due to the economic upheavals following the collapse of communist economies.

The final aspect that I would like to call to your attention is an often-overlooked legacy of the Montreal Protocol: the formation of an international network of hundreds of NGOs, linked by electronic media, that now regularly consult, coordinate positions, and work jointly to influence the course of international environmental issues.

It is hard to believe that as recently as 1985, not a single environmental group attended the signing of the Vienna Convention; and only a handful were here in Montreal ten years ago. The milestone in this development can be dated from Prime Minister Margaret Thatcher's 1989 Conference on Saving the Ozone Layer, which took place several months before the Protocol's first Meeting of Parties, and which was attended by more than ninety environmental NGOs. Gradually expanding in numbers and in degree of collaboration, this network went far beyond traditional green groups to encompass organizations of women, youth, labor, indigenous people, religious groups, and state and local governments. Starting in 1989 with their involvement in the Montreal Protocol, NGOs concerned with the environment evolved into a cohesive force at the 1992 United Nations Conference on Environment and Development in Rio de Janeiro, as well as in the climate, biological diversity and other environmental negotiations.

In conclusion, the Montreal Protocol has been described by U.S. President Ronald Reagan as "a monumental achievement of science and diplomacy", and by the heads of the World Meteorological Organization and UNEP, Dr. Patrick Obasi and Mrs. Elizabeth Dowdeswell, as "one of the great international achievements of the century". Given the extraordinary danger to planetary life that has been avoided and the unprecedented extent of international cooperation that was mobilized, there are few knowledgeable observers that would regard these assessments as hyperbolic. The Montreal Protocol opened a new chapter for the environment by mandating and implementing preventive actions on a global scale—going far beyond national governments—even before actual damage occurred. In the ten years of its existence, the Protocol has become a paradigm for new diplomatic approaches to new international challenges.

Notes

[1] See Margaret Kerr's account in this volume. [Editor's note]

THE MONTREAL PROTOCOL: A NEW LEGAL MODEL FOR COMPLIANCE CONTROL

Patrick Széll[*]

Towards the middle of the Montreal Protocol on Substances that Deplete the Ozone Layer[1] is an unusual provision—unusual in that earlier multilateral environmental agreements contained nothing like it. I refer to Article 8 which provided that:

> The Parties, at their first meeting, shall consider and approve procedures and institutional mechanisms for determining non-compliance with the provisions of this Protocol and for treatment of Parties found to be in non-compliance.

It is fair to say that no-one at the time of negotiation had a very clear idea of what this provision could, or should, lead to, but the Working Group convened to draw up the "procedures and institutional mechanisms" recognised that in the context of ozone layer depletion, prevention has to be preferred to cure. It soon identified a compliance gap between the Protocol's data reporting requirement and its dispute settlement Article that ought to be filled if the treaty was to operate with maximum efficiency.

The compliance regime developed for the Montreal Protocol[2] is frequently spoken of as a major success and in a number of ways it undoubtedly has been. The Parties have agreed to their compliance with the Protocol's obligations being scrutinised by a Committee composed of a cross-section of fellow Parties (Sand 1996, 786-788 and 793). They have accepted that such scrutiny may be triggered by a Party or by the Secretariat. They are prepared for the Committee to ask them searching questions about their performance and to advise them on ways of entering into (or coming back into) full compliance. They have even accepted that the Committee should be able to make recommendations to the Meeting of the Parties on the basis of which the Meeting may adopt decisions which could be critical of them. No multilateral environmental agreement had ever before resulted in such an intrusive compliance control regime.[3] This marks it out as a noteworthy development, as does the fact that the Implementation Committee established to operate the regime has met some seventeen times since 1990[4] and in the process has reviewed a large amount of reported data brought to its attention, principally by the Secretariat.

It is easy, however, to be too sanguine about the achievements of the regime. The harsh fact is that after six years of operation it remains difficult to measure precisely how much of significance has been achieved by the Implementation Committee (Schally 1996, 43-50). Merely checking whether Parties have submitted complete annual reports under Article 7, and goading them into doing so when they have not, is necessary and useful but hardly in itself a demonstration of better compliance. More telling yardsticks have been the Committee's recent sustained efforts to encourage and assist various central and east European countries to fulfil their Protocol obligations. These have met with mixed success, at least where the Russian Federation is concerned.[5] It is important for the credibility and effectiveness of the Protocol that its compliance regime be seen to improve Parties' observance of their obligations. Only when it does so, will the reputation of the regime, of its Committee, and of the Protocol itself be assured.

Notwithstanding the absence of a proven record of delivering significant results, the Montreal Protocol's compliance regime continues to be referred to in environmental circles as an important development that could, with advantage, be copied by other multilateral environmental agreements. Some have taken this advice very literally. For instance, two of the Protocols to the UN/ECE's Convention on Long-Range Transboundary Air Pollution (the Geneva Convention),[6] those concerning the control of emissions of volatile organic compounds (VOCs)[7] and the further reduction of sulphur emissions,[8] contain provision very similar to that of Article 8 of the Montreal Protocol.[9] In the latter case, indeed, a complete compliance regime was negotiated in parallel with the Protocol and the key elements of it were entrenched in the agreement itself (Széll 1994, 104-106). Currently, the Executive Body of the Geneva Convention is studying, by means of a small Expert Group on Implementation,[10] whether, and if so how, the Montreal model could be applied and operated for the benefit of *all* the Convention's Protocols—existing, in the pipeline and yet to be developed. The initial conclusion of the Expert Group was that a single regime could be developed, that it would best take the form of a decision of the Executive Body (not a series of amendments or a new Protocol) and that, subject to countries not being permitted to participate in cases of Protocols to which they are not a Party, there is no reason why a single Implementation Committee should not deal with casework under all the Protocols.[11] The Group's final report was considered by the Executive Body at its fifteenth meeting in December 1997.

The fact that the Parties to the Geneva Convention have already, on two occasions, adopted the Montreal model almost in its entirety is not a reflection of lack of imagination or laziness on their parts. Those involved were well aware of the arguments in favour of tailoring compliance regimes to the particular circumstances of individual environmental agreements (Széll 1994, 108). They found, however, not only that the Montreal model was structurally a sound one but also that the obligations in the two instruments were precise enough to justify applying a very similar process. This contrasts with the Biodiversity Convention[12] and the Desertification Convention,[13] neither of which contains commitments of a sufficiently precise or focused nature to enable a compliance regime to operate meaningfully; and it could well explain why neither of these two treaties includes a provision establishing a compliance regime or requiring one to be developed.

The Basel Convention,[14] too, does not contain provision for a compliance regime, but the reasons for the omission in that case must be different for its procedural requirements are very clear and precise. First, the Basel Convention pre-dates the realisation by states that something more than data reporting requirement and a settlement of disputes Article is required for the effective supervision of compliance with the commitments in multilateral environmental agreements. Secondly, the purpose of the obligations in the Basel Convention is to protect individual Parties from harm rather than to avoid damage to the world as a whole. Thirdly, individual shipments of hazardous wastes that breach the Convention's terms are more likely to result from "malice or greed" (Rummel-Bulska 1996, 53) on the part of an individual trader than from any technical inability by a Party to operate national control measures. Fourthly, in the case of illegal traffic in hazardous wastes, polluter and victim are easier to identify than in the case of, say, depletion of the ozone layer or emissions of sulphur. The Montreal Protocol's compliance regime, or procedures like it, would seem to have little practical relevance to the factual circumstances that the Basel Convention seeks to address. While one can see that, unlike in the case of the Montreal Protocol, there could be a role under the Basel Convention for the operation of a liability regime and how observance of the Convention's requirements might be enhanced by the liability and compensation Protocol that is currently under negotiation, one is left wondering what need, or role, there is for the Convention to have a compliance regime as well (Xueman Wang, n.d.) ; the terms of such a regime are, however, being considered by the Parties in a parallel exercise.

One other recent multilateral environmental agreement contains provision regarding compliance: the Climate Change Convention.[15] It, however, lays down a requirement that, while superficially similar to Article 8 of the Montreal Protocol, is more opaque (see Ott 1996 and Victor 1996b). It provides in Article 13 that the Conference of the Parties:

> shall, at its first session, consider the establishment of a multilateral consultative process, available to Parties on their request, for the resolution of questions regarding the implementation of the Convention.

Three differences between the two Articles should be noted. First, Article 13 calls on the Conference of the Parties merely to "consider" developing a procedure, while the Montreal Protocol spoke in terms of "consider and approve". Secondly, the climate change process has to be of a type "available to Parties on their request" whereas the Montreal Protocol is not so limited. Thirdly, the unambiguous Montreal Protocol notion of "mechanisms for determining non-compliance...and for treatment of Parties found to be in non-compliance" is diluted in Article 13 to a "multilateral consultative process...for the resolution of questions regarding the implementation of the Convention". Of these differences, none of which was accidental, the most significant in practice must be the last one. The climate change negotiators were seemingly determined to avoid the development of a regime that was formal and compelling.

During negotiation of the Climate Change Convention, a sizeable number of countries expressed strong doubts about replicating Article 8 of the Montreal Protocol. The less precise language of Article 13 is a reflection of their wish to see the Convention, if it tackled compliance at all, doing so in a milder manner. It is perhaps surprising to find the climate change negotiators indicating so clearly that the Montreal approach was too tough, given how the Montreal Protocol Parties went out of their way to develop a regime that was not confrontational or punitive in character but rather transparent, co-operative, uncomplicated and facilitative.[16]

The ad hoc Working Group established by the Climate Change Convention Parties to consider and make proposals under Article 13 reflected the nervousness of the Convention's negotiators by taking a slow and cautious approach to the task of developing a multilateral consultative process (MCP). It began by drawing up and circulating a detailed questionnaire[17] to all concerned states, intergovernmental organisations and non-governmental organisations seeking their views on the meaning of the terms used in Article 13 and on the action they considered should be taken under the Article. The replies[18] were synthesised[19] by the Secretariat and are being used as background material by the Group. The Group has also held a workshop at which the experience of other international organisations (such as the International Labor Organization (Romano 1996), the Commission of Human Rights and the Montreal Protocol) in the operation of compliance procedures were explained and discussed.[20] Only in December 1996, at its third session, did the Group get down to substantive discussion. Even then its steps were tentative. A number of countries continued to doubt whether there was need for the Convention to have an MCP at all and there was total refusal to allow any points of agreement or disagreement to be registered.[21]

At the Group's fourth session in February 1997, participants were in a more amenable mood. Progress was greatly helped by two steps taken early on the first day:

1) First, there was agreement that the Group should not complete its exercise until the fourth Conference to the Parties at the earliest. It was apparent that participants wanted to know how the Berlin Mandate negotiators were going to approach the question of compliance under the Protocol before taking final decisions regarding an MCP for the Convention.
2) Secondly, certain Parties announced that they were prepared to moderate their earlier tough line which had been in favour of an MCP that was intrusive, that is to say a regime that would be capable of scrutinising the fulfilment of obligations by individual Parties.

With the entire Group now agreeing that the MCP should be advisory rather than supervisory in character and in the sure knowledge that the exercise was not going to be rushed, it became possible to move more quickly and easily than anticipated from the unfocused general debate that had characterised the third session to the tabling of a document in the form of a negotiating text that highlighted, by means of square brackets, the points of agreement and disagreement among participants.[22]

Despite the large number of square brackets in the text, there is already acceptance within the Group, demonstrated by the debate at the Group's fifth session,[23] that the MCP should operate through a Committee whose task it would be to provide individual Parties, on request (whose is not yet clear), with advice on specific implementation problems that they encounter. One of the Group's main points of convergence has been that the process developed must not duplicate the work of other Convention bodies. It has given expression to this aim principally by agreeing to focus the MCP on helping individual Parties, as opposed to groups of Parties or to the Convention Parties as a whole, both of which clearly fall within the field of operations of the Subsidiary Body on Implementation (SBI).

Many points of detail remain to be resolved, however, when the Group resumes its discussions in June 1998. For example, it is still not at all clear whether the Committee should be a standing or an ad hoc body. Then there are questions such as: Which specific problems should the MCP address? What type of end product should the Committee produce? Should the Committee address itself directly to Parties concerned or act through the SBI and/or the Conference of the Parties? Many of the outstanding issues are of a procedural or legal character and will require mainly careful attention to detail by the Group. Others, however, will be controversial or politically charged, such as the proposal tabled by one developing country towards the end of the fourth session to the effect that a key purpose of the MCP should be to provide assistance to developing country Parties in the form of technical and financial support in compiling and communicating information and identifying financial needs for proposed projects and measures.

It is appropriate that the Working Group on Article 13 should have resisted the temptation to copy the Montreal Protocol regime without question. The advisory approach it is in the throes of developing seeks to complement the essentially aspirational character of the commitments of the Climate ChangeConvention. There is, however, a danger in this, in that the Parties to the Kyoto Protocol may, when the time comes, feel that they should as a matter of course imitate whatever process is drafted for the Convention. The possibility of this happening is not, of course, a reason for opposing or altering the content of the MCP that the Working Group develops for the Convention, but in dealing with the equivalent matter in the context of the new Protocol, it is important that states and others do not merely look over their shoulders at the Convention. Rather, they must be prepared to take as innovative and as individual a line with respect to compliance as the substance of the new Protocol needs in order to be effective (Handl 1997, 30-31 and 48-49).

The Kyoto Protocol will, therefore, provide an important test case of whether, within a single family of multilateral environmental agreements, the Parties are prepared to tailor compliance procedures in such a way that, those members of the family made up of mild and flexible commitments can be addressed by means of an advisory regime, while those family members with precise and onerous commitments can be subject to a more intrusive, supervisory process[24] that is as strong as, or maybe even tougher than, the one conceived ten years ago for the Montreal Protocol.

Notes

[*] The views expressed in this paper are those of the author and do not necessarily represent those of the Government of the United Kingdom.

[1] Adopted in Montreal, 16 September 1987, 26 *International Legal Materials* 1550 (1987).

[2] See Decision II/5 of 27 June 1990 and Decision IV/5 and Annex of 25 November 1992.

[3] For further analysis of the Montreal Protocol's compliance regime and its operation, see Koskenniemi 1992; Lang 1996; Trask 1992; Victor 1995 and 1996a, 58-81; and Széll 1994, 99-103 and 106-109.

[4] The Committee's first meeting was held in Nairobi on 11-12 December 1990 (see UNEP/Ozl.Pro/Imp Com/1 /2 ; its eighteenth meeting took place in Nairobi on 2 and 4 April 1997 (see UNEP/Ozl.Pro/Imp Com/18/3).

[5] See in this regard Decisions VII/15 to 19 concerning, respectively, Poland, Bulgaria, Belarus, the Russian Federation and Ukraine, adopted in Vienna on 7 December 1995, and Decision VIII/22 to 25 concerning, respectively, Latvia, Lithuania, the Czech Republic and the Russian Federation adopted in San José on 25-27 November 1996. For a detailed assessment of the situation, see Werksman 1996; for an analysis of illegal trade under the Montreal Protocol, see Brack 1996, 99-114.

[6] Adopted in Geneva on 13 November 1979, 18 *International Legal Materials* 1442 (1979).

[7] See Article 3.3 of the Protocol concerning the Control of Emissions of Volatile Organic Compounds or their Transboundary Fluxes, adopted in Geneva on 18 November 1991, 31 *International Legal Materials* 586 (1992).

[8] See Article 7 of the Protocol on Further Reduction of Sulphur Emissions, adopted in Oslo on 14 June 1994, 33 *International Legal Materials* 1542 (1994) and the Executive Body's related decision on compliance adopted on the same day (ECE/EB.AIR/38, para. 9, and ECE/EB.AIR/40).

[9] Also, the most recent draft of the UN/ECE Convention on Access to Environmental Information and Public Participation in Environmental Decision Making contains, at Article 14 bis, two alternative formulations for a provision on implementation and compliance. Each would require the Parties, at their first meeting, to establish a procedure and institutional mechanism for determining non-compliance and require the procedure to contain provision for public participation. They differ only in that one of the options goes further than the other in spelling out the details of the procedure and its possible objectives.

[10] Established by the Executive Body in November 1995 ; see ECE/EB.AIR/46, para.45.

[11] See the Expert Group's first report to the Executive Body (EB.AIR/R.102, dated 30 August 1996) and the Executive Body's debate and decision on that report (ECE/EB.AIR/49, paras. 62-64 and Annex III).

[12] Convention on Biological Diversity, adopted in Nairobi, 22 May 1992, 31 *International Legal Materials* 851 (1992).

[13] Convention to Combat Desertification in those Countries Experiencing Serious Drought and/or Desertification, particularly in Africa, adopted in Paris, 17 June 1994.

[14] Basel Convention on the Control of Transboundary Movements of Hazardous Wastes and Their Disposal, adopted in Basel, 22 March 1989, 28 *International Legal Materials* 649 (1989).

[15] United Nations Framework Convention on Climate Change, adopted in New York, 9 May 1992, 31 *International Legal Materials* 851 (1992).

[16] See in this regard para. 23.1 of the Declaration of the European Environment Ministers at the Environment for Europe Conference, held in Lucerne, 28 - 30 April, 1993.

[17] See FCCC/AG13/1995/2, para. 17.

[18] For the responses from Parties and non-Parties, see FCCC/AG13/1996/MISC.1 and Add.1; for the responses from intergovernmental and non-governmental bodies, see FCCC/AG13/1996/MISC.2 and Add.1.

[19] See FCCC/AG13/1996/1.

[20] For the Chairman's informal report on the panel presentation and discussions, see FCCC/AG13/1996/2, paras. 7-9 and Annex.

[21] See the report of the Working Group's third session, FCCC/AG13/1996/4, para. 16 and Annex II.

[22] See the report of the Working Group's fourth session, FCCC/AG13/1997/2, para. 13 and Annex II.

[23] See the report of the Working Group's fifth and final session prior to Kyoto, FCCC/AG13/1997/4.
[24] Among the proposals tabled by the negotiating states is one by the U.S.A. that would require the Conference of the Parties, in drawing up an appropriate compliance regime, to develop "an indicative list of consequences, taking into account the type, degree, and frequency of non-compliance" (see FCCC/AGBM/1997/MISC.1/Add.4, p.16). An earlier proposal by the U.S.A. went further and spoke of some of the consequences being "automatic, while others might be discretionary. Consequences could include, for example:
1) denial of the opportunity to sell tons of carbon equivalent emissions allowed through international emissions trading and/or joint implementation; and
2) loss of voting rights and/or other opportunities to participate in processes under the Protocol". (See FCCC/AGBM/1997/3/Add.1, para 208.5.)

References

Brack, D. 1996. *International Trade and the Montreal Protocol.* London: Royal Institute of International Affairs and Earthscan Publications Ltd.

Handl, G. 1997. "Compliance Control Mechanisms and International Environmental Obligations." *Tulane Journal of International and Comparative Law.* 5: 29-49.

Koskenniemi, M. 1992. "Breach of Treaty or Non-Compliance? Reflections on the Enforcement of the Montreal Protocol." *Yearbook of International Environmental Law.* 3: 123-62.

Lang, W. 1996. "Compliance-Control in International Environmental Law: Institutional Necessities." *Heidelberg Journal of International Law.* 56: 685-95.

Ott, H. E. 1996. "Elements of a Supervisory Procedure for the Climate Regime." *Heidelberg Journal of International Law.* 56: 732-49.

Romano, C. P. R. 1996. *The ILO System of Supervision and Compliance Control: A Review and Lessons for Multilateral Environmental Agreements.* ER-96-1: International Institute for Applied Systems Analysis.

Rummel-Bulska, I. 1996. "Implementation Control: Non-Compliance Procedure and Dispute Settlement: From Montreal to Basel." In *The Ozone Treaties and Their Influence on the Building of International Environmental Regimes.* Austrian Foreign Policy Documentation. Austrian Ministry of Foreign Affairs.

Sand, P. H. 1996. "Institution Building to Assist Compliance with International Environmental Law: Perspectives." *Heidelberg Journal of International Law.* 56: 774-795.

Schally, H. M. 1996. "The Role and Importance of Implementation Monitoring and Non-Compliance in International Environmental Regimes." In *The Ozone Treaties and Their Influence on the Building of International Environmental Regimes.* Austrian Foreign Policy Documentation. Austrian Ministry of Foreign Affairs.

Széll, P. J. 1994. "The Development of Multilateral Mechanisms for Monitoring Compliance." In *Sustainable Development and International Law.* London: Graham & Trotman.

———. 1996. "Implementation Control: Non-Compliance Procedure and Dispute Settlement in the Ozone Regime." In *The Ozone Treaties and Their Influence on the Building of International Environmental Regimes.* Austrian Foreign Policy Documentation. Austrian Ministry of Foreign Affairs.

Trask, J. 1992. "Montreal Protocol Non-Compliance Procedure: The Best Approach to Resolving International Environmental Disputes?" *Georgetown Law J.* 80: 1973-2001.

Victor, D. G. 1995a. "The Early Operation and Effectiveness of the Montreal Protocol Implementation Committee." In *Project on Implementation and Effectiveness of International Environmental Commitments for the International Institute for Applied Systems Analysis.* Laxenburg (Austria).

———. 1995b. *Design Options for Article 13 of the Framework Convention on Climate Change: Lessons from the GATT Dispute Panel System.* Executive Report ER-95-1. International Institute for Applied Systems Analysis.

———. 1996. "The Montreal Protocol's Non-Compliance Procedure: Lessons for Making Other International Environmental Regimes More Effective." In *The Ozone Treaties and Their Influence on the Building of International Environmental Regimes.* Austrian Foreign Policy Documentation.

Austrian Ministry of Foreign Affairs.
Werksman, J. 1996. "Compliance and Transition: Russia's Non-Compliance Tests the Ozone Regime." *Heidelberg Journal of International Law.* 56: 750-753.
Xueman Wang. n.d. "Towards a Compliance Regime in Environmental Treaties" (unpublished paper)

THE USE OF TRADE MEASURES IN THE MONTREAL PROTOCOL

Duncan Brack

The 1987 Montreal Protocol on Substances that Deplete the Ozone Layer is one of the great success stories of international environmental diplomacy in the 1980s and 1990s. In sharp comparison to many seemingly intractable problems of pollution and resource depletion, the Protocol's gradually evolving control schedules have proved highly successful in controlling the production and consumption of ozone-depleting substances. While there are many reasons for this outcome, this paper will focus on just one: the incorporation of trade restrictions as a compliance and enforcement mechanism within the Protocol itself. It will then draw conclusions as to the applicability of similar trade measures to other international environmental treaties.

Multilateral environmental agreements and trade measures

The Montreal Protocol is one among over 180 multilateral environmental agreements (MEAs) currently in existence. Over twenty incorporate trade measures. These include three of the most important: the Protocol itself, the 1973 Convention on International Trade in Endangered Species of Wild Fauna and Flora (CITES) and the 1989 Basel Convention on the Control of Transboundary Movements of Hazardous Wastes and their Disposal. Almost all of the remainder concerns the protection of various categories of animals and plants. In the absence of any comprehensive framework of global environmental law, the negotiation of further MEAs—such as the protocol to the Climate Change Convention adopted in December 1997—form an increasingly prominent part of the international agenda, and some of these may well contain trade measures.

Trade provisions in MEAs have been designed to realize four major objectives:

1) To restrict markets for environmentally hazardous products or goods produced unsustainably.
2) To increase the coverage of the agreement's provisions by encouraging governments to join and/or comply with the MEA.
3) To prevent free-riding (where non-participants enjoy the advantages of the MEA without incurring its costs) by encouraging governments to join and/or comply with the MEA.
4) To ensure the MEA's effectiveness by preventing leakage—the situation where non-participants increase their emissions, or other unsustainable behavior, as a result of the control measures taken by signatories.

While CITES and the Basel Convention are MEAs aimed explicitly at controlling trade—the first and second objectives listed above—the Montreal Protocol employs trade measures as one policy instrument among several in achieving its aims. The trade measures are there to realize objectives three and four: the control of free-riding and of leakage.

Trade measures and the Montreal Protocol

Article 4 of the Protocol imposes bans on trade between parties and non-parties to the treaty. They cover restrictions on both imports from and exports to non-parties of three types of products: ozone-depleting substances (ODS); products containing ODS (e.g. refrigerators); products made with but not containing ODS (e.g. electronic components).

Originally applied to Annex A ODS (the main chlorofluorocarbons (CFCs) and halons), these measures have been gradually extended to the other categories of ODS, with their application to Annex E (methyl bromide) which was under consideration in 1997 (see Table 1). Annex D of the Protocol lists the categories of products identified as "products containing" Annex A ODS: automobile and truck air conditioning units; domestic and commercial refrigeration; air conditioning and heat pump equipment; aerosol products (except medical aerosols); portable fire extinguishers; insulation boards, panels and pipe covers; and pre-polymers. Similar listings have not been carried out for other categories (Annex B and Annex C group II) since phase-out of the ODS themselves rendered them unnecessary.

The category of "products made with but not containing" ODS originally represented a potentially large number and variety of goods—most electronic components, for example, were treated with CFC-containing cleaning and degreasing agents. The Parties thus agreed simply to "determine the feasibility" of applying trade restrictions to such products within five years of the Protocol's entry into force. When this was discussed, in 1993, the Parties decided that the problems of detection were so large, the range of products which would have to be tested was so wide, and the quantities of ODS which were involved were so small, that these trade restrictions should not be implemented, though the issue was to be reviewed at regular intervals. This is an important issue in the case of methyl bromide, since many of the products treated with the chemical are traded.

Table 1. Trade restrictions between parties and non-parties (Montreal Protocol as amended and adjusted by the London and Copenhagen Meetings)

Date of trade ban with non-parties in:	ODS Annex:			
	A (CFCs, halons)	B (CFCs, MCI, CTC)	C (HCFCs)	E (MBr)
Imports of ODS	January 1990	August 1993	June 1995	June 1999
Exports of ODS	January 1993	*August 1993*	June 1995	June 1999
Imports of goods Containing ODS:				
Listing	January 1992	(August 1995)	(June 1997)	June 2001+
Ban	January 1993	(August 1996)	(June 1998)	June 2002+
Imports of goods made With but not containing ODS:				
"Determination of feasibility"	January 1994	August 1997	June 1999	June 2003

Notes: (a) Dates in parentheses relate to restrictions which were not implemented. (b) The extension of the trade measures to Annex C Group I ODS (hydrochlorofluorocarbons or HCFCs) was due to be discussed at the 1995 Vienna Meeting but in fact was not. (c) The extension of the trade measures to Annex E ODS (methyl bromide) was discussed at the 1997 Montreal Meeting. The dates in italics show the outcome of the current proposals for adjustment of the Protocol, assuming it comes into force nine months after it is agreed. There would be an additional delay for imports of goods containing methyl bromide, as there is an additional "determination of feasibility" stage proposed.

The negotiators of the Montreal Protocol had two aims in drawing up these trade provisions. One aim was to maximize participation in the Protocol by shutting off non-signatories from supplies of CFCs and providing a significant incentive to join. If completely effective, this would in practice render the trade provisions redundant, as there would be no non-parties against which to apply them.

The other goal, should participation not prove total, was to prevent industries from migrating to non-signatory countries to escape the phase-out schedules. In the absence of trade restrictions, not only could this reaction fatally undermine the control measures, it would also help non-signatory countries gain a competitive advantage over signatories, as the progressive phase-outs raised industrial production costs. If trade were forbidden, however, non-signatories would not only be unable to export ODS, they would also be unable to enjoy fully the potential gains from cheaper production as exports of products containing, and eventually made with, ODS, would also be restricted. (In fact, as industrial innovation proceeded far more quickly than expected, many of the CFC substitutes proved significantly cheaper than the original ODS—but this could not have been foreseen in 1987.)

The evidence suggests that the trade provisions achieved their objectives. All CFC-producing countries and all but a handful of consuming nations have adhered to the treaty. Although it is difficult to determine states' precise motivations for joining—there are a variety of reasons, including the availability of financial support for developing countries—the trade restrictions do appear to have provided a powerful incentive, and at least some countries have cited them as the major justification for their participation.

The Republic of Korea offers the clearest case. Korea delayed accession to the Protocol until 1992, reportedly because it believed that the resources available under Multilateral Fund assistance would have been inadequate to cover the adaptation costs of its large and growing electronics industry. Korean domestic ODS production increased in the early years of the Protocol (from thirty-six percent of consumption in 1989 to fifty-two percent in 1990, against a thirty percent fall in total consumption) and the country almost certainly could have reached self-sufficiency. Nevertheless Korea eventually acceded; the threat of trade restrictions by Montreal Protocol member countries (such as the United States or the European Union) against Korean exports such as refrigerators or air conditioning systems (particularly in cars), or, potentially even worse, on the many Korean electronic goods produced with but not containing ODS (which was still at the time—and in theory still is—possible ground for trade measures), proved a powerful incentive. Similar considerations may have influenced the other industrializing countries of South-East Asia, and the threat of trade bans played a role in Israel's decision to ratify the Protocol in 1992.

Similarly, Mexico, with significant investment in electronics plants along its border with the U.S.A., was one of the few developing countries to join the Montreal Protocol at the outset. Most other developing countries acceded after the treaty was amended, in 1990, to a total phase-out of CFCs and halons. However small a country's production and consumption, it would then eventually be cut off from sources of supply and/or face restrictions on its exports if it remained outside the Protocol. And the more countries that joined, the greater the incentive for the rest to join, particularly if there were no significant producers left outside; the trade provisions would then shut off consumers from any legal source of supply.

The major CFC producers, mostly located in the U.S. and Western Europe, and therefore subject to the controls from the start, were supporters of the trade restrictions, viewing them as a method of ensuring that the alternatives to CFCs they produced were not undercut by cheaper competition from non-parties.

Finally, there is no evidence to suggest that industrial migration has occurred to any significant extent, although, of course, there are now so few non-parties that this would be almost impossible. It remained a matter of some concern as to whether ODS-producing industries would move their production facilities to developing countries, with their longer phase-out periods, as phase-out approached in their home countries. The United Nations Environment Programme (UNEP) and the United Nations Conference on Trade And Development (UNCTAD) studies, however, have failed to find any significant evidence supporting this fear.

Trade measures in other MEAs

It is clear, then, that the trade provisions of the Montreal Protocol were a vital component in (a) building the wide international coverage it has achieved and (b) preventing industrial migration to non-parties in order to escape the controls on ODS.

In principle, similar provisions may form an important component of future MEAs that aim to place limits on the production and consumption of polluting substances or on behavior. This applies to MEAs, which, like the Montreal Protocol, may use trade measures as an enforcement and compliance mechanism alongside other policy instruments—phase-out schedules, financial and technology transfer, etc.—in achieving their aims. The Kyoto Protocol to the U.N. Framework Convention on Climate Change and any agreement on the control of persistent organic pollutants, are potential examples. MEAs which are aimed explicitly at controlling trade, such as CITES and the Basel Convention, raise somewhat different issues, since the trade measures are the whole point of these treaties—but much of the comments below are still relevant.

Three sets of issues need to be examined when considering the value of such trade measures in any given MEA: feasibility, fairness, and interrelationship with the multilateral trading system. They are considered below, particularly in relation to the Climate Change Convention, since these issues have been raised in the debates around it.

Feasibility

Montreal Protocol-type trade measures, i.e. bans on trade, will be easiest to implement if the products being controlled are:
1) limited in type and application;
2) limited in origin;
3) easily detectable; and
4) easily substitutable.

The main (Annex A) CFCs, for example, fitted this bill quite well. Only a limited number of countries and producers manufactured them. Although this range widens when products containing CFCs are considered, it still forms not a very substantial proportion of world trade, or of any single country's exports and imports. Note that the limits are much looser when products made with but not containing CFCs are added—which is one of the reasons why the parties decided not to implement this section of the Protocol (though the volume of CFCs which would have been affected was also quite limited). CFCs can be hidden, and are difficult to detect, requiring, for example, chemical analysis to prove their presence if their containers are mislabelled. Indeed, illegal trade is a growing problem for the implementation of the Protocol, but the steps taken by the U.S.—and, recently, E.U.—authorities to tackle the problem show that it can be countered quite effectively. Finally, CFCs have turned out to be relatively easily substitutable—one of the main reasons why the Protocol has proved to be such a success—so any argument against trade measures resting on the dependence of any country on trade in CFCs is not strong. For all these reasons, trade measures in the Protocol are both technically and politically feasible.

Rather greater problems are encountered in applying these tests to the case of climate change. Greenhouse gases, of course, are not precisely analogous to ODS, since most of them—carbon dioxide, some sources of methane, nitrous oxide—are essentially byproducts, whereas others—HFCs, some perfluorocarbons (PFCs)—are themselves products which are traded. For this latter group, Montreal Protocol-type trade restrictions could be applied, and the issues and problems are very similar.

For the former, and much more important, category, trade restrictions would have to be applied against either the commodities which release greenhouse gases when used—i.e. fossil fuels—and/or goods made with processes which release greenhouse gases, which in practice means the vast majority of manufactured goods. These largely fail all the four tests set out above: they are not limited in type or origin, they are not easily detectable, and they are not easily substitutable. However, this case is not quite the same as the Montreal Protocol. Since ultimate phase-out of greenhouse gases is not the aim, total trade bans might not be so appropriate; duties or taxes, for example, could be applied against various categories of imports from non-parties. Implementation could be tied to reciprocal obligations on the part of the parties—such as the achievement of particular greenhouse gas reduction targets, and/or the removal of trade-distorting and climate change-accelerating domestic policies, such as agricultural protection or subsidies for energy industries.

Even under these circumstances, if such trade measures were agreed and employed, they would represent a very severe restriction on trade, and an accompanying high welfare loss. By the same token, however, they would create a massive incentive to join and adhere to the agreement, which, after all, is the point of the trade measures. The Montreal Protocol trade measures have in fact hardly ever been used, since almost every country is now a Party to the treaty.

Fairness

If the trade measures are effective in persuading or compelling countries to ratify the MEA in question, it is important that the agreement be fair. This argument has two aspects: whether the agreement itself is firmly founded on good science and a full appreciation of the issues; and whether its provisions are equitable for countries of different types, e.g. industrialized and developing. It is clearly undesirable, from the point of view of building an effective international environmental regime, if the threat or use of trade measures compels a country to join an MEA which it regards as unnecessary and which imposes excessive costs on its economy.

Once again the Montreal Protocol scores relatively highly. Although there may have been some doubts over the science and impacts of ozone depletion in the early years, there are no serious ones now. The differential phase-out schedules applied to non-Article Five and Article Five parties, and the establishment of the Multilateral Fund, with its explicit aim of meeting the "incremental costs" of Article Five parties in complying with the requirements of the Protocol, meet the equity point. If the Fund works as it is intended, developing countries are no worse off as signatories than they are as non-signatories—and, indeed, they should be better off, since their adherence to the Protocol reduces the total extent and costs of ozone depletion.

This time the climate change regime also scores relatively well. The science of climate change is perhaps as good now as was the science of ozone depletion in the mid-1980s, but, clearly, the more credible it is, the stronger will be the agreement. The Framework Convention already treats industrialized ("Annex I") countries differently from developing countries. Under Article 4.2, the former group is to "take the lead in modifying longer-term trends in anthropogenic emissions", and under 4.3, Annex II parties (basically, the Organization for Economic Cooperation and Development (OECD)) are to provide financial support for developing country commitments, i.e. reporting requirements. It is inconceivable that emission limitation and reduction objectives for developing countries would ever be agreed without an effective financial mechanism, whether organized through the Global Environment Facility or through some new institution. The idea of joint implementation included in the Kyoto Protocol already provides a possible route for finance and technology transfer to developing countries.

Interrelationship with the multilateral trading system

Any MEA trade measures are likely in principle to be inconsistent with the General Agreement on Tariffs and Trade, the GATT. As essentially a proscriptive agreement, defining what contracting parties may not do (or may do under certain circumstances), interpretation of the GATT proceeds through a case law-type approach, following rulings by dispute panels in particular cases. Although there has not yet been such a case involving an MEA, the reasoning used by GATT and World Trade Organization (WTO) dispute panels in a number of trade-environment cases suggests that MEA trade provisions might well be found to be in breach of various provisions of the GATT if such a challenge were ever to be brought. The problem is, of course, that trade measures are designed specifically to discriminate between countries based on their membership of the MEA or their environmental performance, whereas the essential basis of the GATT is to prevent discrimination in trade.

This topic in fact became one of the most important items of debate in the WTO's Committee on Trade and the Environment in preparing its first report for the WTO Ministerial meeting in December 1996. Discussion saw members putting forward proposals designed variously to define under what conditions trade measures taken pursuant to an MEA could be considered to be "necessary" under the terms of GATT's Article XX (the "general exceptions" article), or to establish a degree of WTO oversight on the negotiation and operation of trade provisions in future MEAs. No consensus was reached about the need for modifications to trade rules.

Many proposals have been made for the resolution of this potential clash between international regimes, though it is difficult to see how the WTO is likely to agree on any of them in the foreseeable future. The extent to which this is regarded as a problem usually depends on whether one regards trade liberalization or environmental protection as the superior objective. But, as above, if the trade measures were to be successful in their aim of persuading countries to join, they would not be used, and the problem therefore disappears.

Conclusion

I have argued above that the trade measures of the Montreal Protocol were a vital element in the success of the agreement, and that in principle similar provisions may have an important role to play in other MEAs. Their precise form will of course vary with the MEA in question, and in some cases they are likely to be more politically credible and technically feasible than in others. (They may be more easily applicable, for example, to an agreement controlling persistent organic pollutants than to the Climate Change Convention.) One other general conclusion is that they should always be accompanied by effective finance and technology transfer mechanisms if the MEA is to be regarded as fair.

Ideally, the presence of trade measures should provide a sufficient incentive to result in universal participation—a conclusion which is supported by theoretical modeling suggesting that MEAs with trade restrictions evolve to one of two self-enforcing equilibrium points: universal participation or zero participation. Since, in general, modern governments are not run by theoretical modelers, the enforcement of any trade measures written into an MEA must at least be contemplated. In addition to matters of technical feasibility, the question of the interrelationship with the multilateral trading system, therefore, must also be considered.

In purely legal terms, this could be met within the WTO through an amendment to the GATT, a position for which there is some support (including from the European Union), though also much opposition. In political terms, it is a question of both whether and how MEAs can be enforced against non-participating countries. If the objectives of the MEA are accepted as valid, and if the actions of non-participants inflict physical damage on the members of the agreement—which, in the case of transboundary or global pollution, they always do—then a strong case can be made for discriminatory measures directed against non-participants. Furthermore, there are a limited number of routes by which countries can affect the actions of other countries: diplomatic pressure, provision of financial and technological assistance, trade sanctions and military force. While the first two are clearly preferable, they have obvious limits. Trade measures are likely to continue to play a role as one component of effective environmental agreements.

THE MONTREAL REGIME: STICKS AND CARROTS

COMMENTS

Peter H. Sand

Your commentator has been admonished, above all, to be critical. That puts me in a very delicate position—for it so happens that I find nothing to disagree with in the papers of our two panelists, for whose professionalism I have the greatest admiration. If anything, it probably is that innate British sense of understatement, which seems to lead both of them to err on the side of caution rather than on the side of grand "new legal models".

Yet, successful models have a way of developing their own dynamics, irrespective of parental guidance. For example, Patrick Széll cautiously speculates that there really is nothing to commend the Montreal Protocol as a model for the 1994 Convention on Desertification; little does he seem to realize that the Desertification Secretariat has already embarked on the preparation of "procedures to resolve questions of implementation" under article 27 of the Convention, drawing heavily on Patrick's pioneering work with the Montreal and Geneva procedures.[1] With his usual modesty, he also fails to mention that the Lucerne Conference of European Environment Ministers in 1993 called for the development of "non-confrontational" compliance procedures (*à la Montréal*) for *all* multilateral environmental agreements[2]—which is why such procedures are now popping up in the new Economic Commission for Europe (ECE) draft convention on environmental information and public participation. So the "Montreal model" keeps multiplying and cloning itself; and I sometimes picture Patrick as the sorcerer's apprentice who started it all, and who would now wish to reach for his magic wand to get those genies back into the bottle. We may indeed be witnessing a process of inter-organizational "social learning" (Haas and Haas 1995) not unlike the transcultural spread of innovations, which Arnold Toynbee described as *mimesis* (imitation), i.e., "the reception and adoption of elements of culture that have been created elsewhere and have reached the recipients by a process of diffusion." (Toynbee 1961, 343 and Gabor 1971)

Let us start by clarifying our terminology, though. The papers by Patrick Széll and Duncan Brack really are not about the Montreal Protocol at all—at least not about the 1987 text which we are celebrating today. What we are talking about is the "Montreal regime", as it subsequently evolved and continues to evolve, through amendments, adjustments and interpretations; and here I deliberately do not use precise legal terms, but the jargon of political scientists, who tell us that international regimes are "social institutions consisting of agreed-upon principles, norms, rules, procedures and programs that govern the interaction of actors in specific issue-areas." (Levy, Young, and Zürn 1995)

The distinction is less academic than it may sound. The fact that the Montreal Protocol is not just a 1987 treaty, but—as Winfried Lang put it—"a living organism", also has important consequences for the trade measures discussed by Duncan Brack: the issue of compatibility with the General Agreement on Tariffs and Trade or GATT (the world trade regime) arises not merely with regard to the original treaty rules, but typically with regard to trade restrictions that may subsequently emerge from a treaty regime, through dynamic application and interpretation by the Conference of Parties to the treaty. An illustrative example is the 1973 Convention on International Trade in Endangered Species of Wild Fauna and Flora (CITES) (United Nations 1976, 243) the actual trade sanctions of which—against Parties and non-Parties!—were developed only after the treaty had been negotiated and adopted.[3] I can assure you that the World Trade Organization is rather more worried about this secondary, "regime-born", type of trade barriers than about the original trade-related provisions found in the text of multilateral environmental agreements such as CITES and the Montreal Protocol.

By the same token, we must not forget that the "non-compliance procedure" (which is the focus of this panel) is part and parcel of other regime features that were also developed well after 1987, and that were also grafted upon the original text by skillful innovators: in particular, the procedure for compliance assistance—especially the Multilateral Fund and its concept of "incremental costs", under which designated developing countries are entitled to obtain substantial financial assistance to meet their treaty obligations.[4]

In other words, the "Montreal regime" as it exists today is a package deal: it not only wields the innovative "stick" of the non-compliance procedure explained by Patrick Széll, but also the innovative "carrot" of incremental cost subsidies, which in turn have become a model for other international regimes and the Climate Change Convention in particular.[5] What is crucially important here is that we are dealing with a combination of sticks and carrots. Suffice it to say that this package deal was even subsequently extended to grant "carrots" to a group of countries not originally entitled to them; i.e., the East European economies in transition, which Patrick Széll and Duncan Brack rightly identify as major problem cases.[6]

Should you look for that important modification in the text of the Montreal Protocol, you will—once again—be disappointed: it was arranged in 1991 between the World Bank and the Executive Committee of the Montreal Protocol Multilateral Fund, in the context of what has since become known as the "GEF", the Global Environment Facility.[7] There is not a word about it in any protocol amendment, but the GEF Operational Strategy says that "the GEF will assist otherwise eligible recipient countries that are not Article 5 countries (i.e., designated developing countries), or whose activities, while consistent with the objectives of the Montreal Protocol, are of a type not covered by the Multilateral Fund;" provided, however, that they are Parties to the Protocol, have ratified the London amendments, and are in compliance with their obligations under the amended Protocol.[8]

I am afraid that is where the innovative Montreal package begins to raise some novel legal problems, too; especially when it comes to reconciling the regime with certain general principles of international law. The problems are basically two-fold: First, if we start from the traditional doctrine of equality of sovereign states, and their common duty to comply with treaty obligations in good faith, some commentators have indeed found the idea of "subsidizing compliance" for a selected few Parties downright outrageous—a travesty, as it were, of the time-honored principle of *pacta sunt servanda.*[9]

Let us be realistic, though: we all know that neither China nor India would have joined the Montreal regime, had it not been for the prospect of multilateral funding to phase out their chlorofluorocarbon (CFC) production. International lawyers therefore were quick to seek and find a plausible rationale for this type of side-payment (at least with regard to designated developing countries): *viz.*, the new concept of "common but differentiated responsibilities",[10] which explains why some countries should pay to help others to comply. With that equitable explanation, the alarmed champions of *pacta sunt servanda* may be able to sleep in peace again.

There is another problem with our Montreal package deal, however, which goes to the heart of the non-compliance procedure as it is now evolving. The Implementation Committee (whose accomplishments both Patrick Széll and Duncan Brack acknowledge) typically evaluates compliance by "problem countries", and has repeatedly had occasion to do so with regard to Eastern Europe. Obviously, a harsh finding of non-compliance would deprive those countries of the "carrot" of GEF funding; and so the Implementation Committee tends to lean over backwards to find them more or less in compliance—possibly in order to keep their goodwill, and not to lose them as treaty Parties altogether.

It is true that such pragmatic interpretation keeps the treaty flexible. In affirming their power to make their own evaluation of what constitutes compliance, the Contracting Parties may (arguably) even grant exemptions from the strict application of treaty rules. Yet, consensual redefinition of treaty standards—however well-meaning, albeit for the sake of avoiding conflict—also tends to "soften" the entire treaty regime, and thereby risks to weaken its effectiveness in the long run.[11] What that reminds me of is a subtle Italian way of describing justice in Sicily, the land of the Godfather:

La legge é applicata al némico—ma interpretata all'amico.
[Law is applied to the enemy—but interpreted to a friend.]
Perhaps my concerns are old-fashioned (or do I have a cultural problem there?); but I will not hesitate to confess that I have serious difficulty with that kind of pragmatism.

Notes

[1] See U.N. doc. A/AC.241/50 (1995), and the report of the eighth session of the Intergovernmental Negotiating Committee on Desertification (Geneva, February 1996). On the systematic transfer of "lessons learned" from the Montreal Protocol to the 1991 and 1994 protocols of the Geneva Convention on Long-Range Transboundary Air Pollution, see Széll 1995, 97-109.
[2] Lucerne Declaration, section 23(1); see *Yearbook of International Environmental Law* 4 (1993), doc. 9.
[3] For case histories see Sand 1997, 38-40.
[4] Pursuant to the amendments of the Protocol adopted by decision II/8 at the 1990 London meeting, *International Legal Materials* 30 (1991) 537; see DeSombre and Kauffman 1996, 89-126.
[5] On these incentives and disincentives, see Sand 1996a.
[6] On illegal trade, see Brack 1996, 99-114.
[7] Restructured in 1994, *International Legal Materials* 33 (1994) 1283; see Boisson de Chazournes 1995 and Sand 1996b, 479-499.
[8] GEF *Operational Strategy* (Washington/DC 1996), p. 64. This "complementary" GEF project assistance is distinct from the World Bank's management of Multilateral Fund projects for Article 5 countries, pursuant to the "Ozone Projects Agreement" concluded with the Executive Committee on 9 July 1991, Supplement to Annex D of World Bank Resolution 91-5 of 14 March 1991, *International Legal Materials* 30 (1991) 1773.
[9] For warnings against the risk of "undermining the credibility of international environmental law", see the summary of discussions at the 1996 Workshop on Institution-Building in International Environmental Law, *Heidelberg Journal of International Law* 56 (1996) 820, at 827; see also the *caveat* against "subsidized compliance" in U. Beyerlin & T. Marauhn, *Law-Making and Law Enforcement in International Environmental Law after the 1992 Rio Conference*, UBA-Berichte 97: 3 (Berlin 1997).
[10] Principle 7 of the Rio Declaration on Environment and Development, which may be seen as reflecting article 5(5) of the Montreal Protocol (as amended in 1990), article 4(7) of the Climate Change Convention, and article 20(4) of the Biodiversity Convention.
[11] See also Koskenniemi 1996, 236, at 248 (quoting Sir Robert Jennings), as to the continuing need for objective ascertainment of a breach of international treaty law.

References

Boisson de Chazournes, L. 1995. "Le Fonds pour l'environnement mondial: recherche et conquête de son identité." *Annuaire Français de Droit International.* 41: 612-632.
Brack, D. 1996. *International Trade and the Montreal Protocol.* London: Royal Institute of International Affairs and Earthscan Publications Ltd.
DeSombre, E. R. and J. Kauffman. 1996. "The Montreal Protocol Multilateral Fund: Partial Success Story." In *Institutions for Environmental Aid: Pitfalls and Promise.* Eds. Robert O. Keohane and Marc A. Levy. Cambridge: M.I.T. Press: 89-126.
Gabor, D. 1970. *Innovations: Scientific, Technological and Social.* London: Oxford University Press.
Haas P. M. and E. B. Haas. 1995. "Learning to Learn: Improving International Governance." *Global Governance.* 1: 255-285.
Koskenniemi, M. 1996. "New Institutions and Procedures for Implementation Control and Reaction." In *Greening International Institutions.* London: FIELD/Earthscan.
Levy, M. A., O.R. Young and M. Zürn. 1995. "The Study of International Regimes." *European Journal of International Relations.* 1: 267-330.

Sand, P. H. 1996a. "International Economic Instruments for Sustainable Development: Sticks, Carrots, and Games." *Indian Journal of International Law.* 36(2): 1-16.

———. 1996b. "The Potential Impact of the Global Environment Facility of the World Bank, UNDP and UNEP." In *Enforcing Environmental Standards: Economic Mechanisms as Viable Means?* Ed. Rüdiger Wolfrum. Berlin: Springer.

———. 1997. "Whither CITES? The Evolution of a Treaty Regime in the Borderland of Trade and Environment." *European Journal of International Law.* 8: 29-58.

Széll, P. 1995. "The Development of Multilateral Mechanisms for Monitoring Compliance." In *Sustainable Development and International Law.* Ed. Winfried Lang. London: Graham & Trotman.

Toynbee, A. J. 1961. *A Study of History : Reconsiderations.* Vol.12. London: Oxford University Press.

United Nations Treaty Series. 1976. No. 993.

Werksman, J. 1996. "Compliance and Transition: Russia's Non-Compliance Tests the Ozone Regime." *Heidelberg Journal of International Law.* 56: 750-773.

INTERNATIONAL COOPERATION
An example of success

Juan Antonio Mateos

International cooperation has served to promote political or economic interests for many decades. On the whole, it has brought little benefits to the societies at the receiving end. During the long period of the Cold War, the provision of cooperation was tied to military and strategic alliances in the context of the struggle between the two giants of that time. International cooperation, whether bilateral or multilateral, has been inconsistent and devoid of visible results for the political, economic and social development of recipient countries. Many African and some Central American countries are live examples of this lack of success.

International cooperation has also occupied and important place in the North-South confrontation. The efforts of the developing countries to obtain financial commitments from the developed countries have been ignored. Not only has the goal, established two decades ago, that developed countries dedicate 0.7 percent of their gross internal product to "official assistance for development" not been reached, but the available figures for cooperation indicate that these resources have been constantly diminishing.

Towards the end of the eighties, the international agenda began to change. Items which were considered very marginal priorities moved up. Most prominent was the environment which represented a clear consensus of the international community and, therefore, a real possibility of experimenting new forms of international cooperation.

Even though the question of the environment had been taken up since the seventies, it enjoyed limited attention. Paradoxically, this lack of attention gave those involved in this issue the liberty to continue elaborating a solid international legal framework, consistent with scientific evidence and the political interests of the states to promote research, human resources training and support experimental projects. At the international level, the United Nations Environment Programme

(UNEP) kept the focus on environmental issues. In the process of elaborating an international environmental legal framework, the negotiations, which made possible the Montreal Protocol, were concluded in 1987 without having drawn major comments from the media. Governments considered the new Protocol a marginal event compared to the issues that occupied the attention of the international community at the time.

Nevertheless, the last three years of the eighties witnessed spectacular transformations of the international forces. The Cold War came to an end and the international agenda was modified. New winds blew in favor of international cooperation raising large expectations around the resources that might be made available for this purpose. New horizons were contemplated with generosity. The language of cooperation replaced that of confrontation. It was in this context that amendments to the Montreal Protocol were negotiated, giving rise to a particularly successful instance of cooperation between developed and developing countries.

Which were the elements that allowed for a successful negotiation?

1) Scientific evidence. It was fully proven at that time that well-identified anthropogenic substances were the cause of the depletion of the stratospheric ozone layer. Scientific work such as that of Dr. Mario Molina, which would later be fully recognized, entirely supported the decision of reducing and, eventually, eliminating the production and consumption of substances controlled by the Protocol.
2) There had also been proof of the existence of a "hole" in the ozone layer in the Antarctic region. It was also known that the depletion of such layer could provoke diverse diseases and affect cultivation and livestock. In this context the word "cancer" was a very important detonator.
3) The awareness that reverting the process of depletion required the participation of all countries, producers as well as consumers of substances controlled by the Montreal Protocol, regardless of their level of development.
4) The political will of States to create new bases for international cooperation in order to confront what had begun to be called global issues. Their global character required a global solution.
5) The recognition by the so-called "developed countries" of their responsibility in the depletion of the ozone layer.
6) The acceptance by the firms producing or using the controlled substances that the Protocol's provisions were irreversible.
7) The readiness of negotiators to face the challenges with a certain imagination.

The spirit of cooperation prevailed over the demons of confrontation. Compromise formulas hitherto unthinkable were accepted. A Fund was accordingly created by which developed countries would provide the resources necessary for developing countries to comply with the Protocol. The principle of One State One Vote was accepted along with equal representation of both groups of countries in the Executive Committee of the Fund. At last, a relationship among equals existed.

Allow me an anecdote that reflects the atmosphere prevailing during the 1990 London meeting where the Montreal Protocol was amended. On a not very cold early morning, some of the main problems were discussed with vigor by a few of us. It was absolutely necessary to finish that same evening. As the discussion progressed, throats went dry and ideas stopped flowing with their usual intensity. All of us tried to resist the density of smoke from the cigarettes which some smoked with great abandon (fundamentalists were still few then). Thirst threatened to frustrate our efforts. At three o'clock in the morning, in London, in an office of an international organization, the only impossible thing was to get a little liquid to refresh throats and ideas. Two Ambassadors, one from a developed country and one form a developing country, decided to organize a small expedition in search of the much-desired liquid. In fact, the Ambassador from the developed country was charged with going into the offices of the executive director of UNEP and the Ambassador from the developing country with exploring the possibilities of finding a neighborhood bar. This mission of cooperation was successful. There was liquid for all the participants late in the meeting. On the following day, Dr. Tolba found a handwritten message in his office that said: "Please, replenish the bar".

It was this atmosphere of cooperation and understanding that made possible an agreement which remains historical. There was commitment and generosity on both sides of the table and there was also the presence of an individual whose energy stimulated everyone: Dr. Mostafa Tolba. I would almost dare to say that the euphoria of the moment saw even the most reluctant agreed, only to regret it soon thereafter.

This achievement was soon the target of criticisms and attempts to reduce its importance. As a matter of fact, some of the developed countries that were not included in the first Executive Council expressed all kind of doubts regarding the experiment's viability. Representatives of this group of countries in the Council attempted to incorporate control measures and new restrictions in order to make the disbursements of funds more difficult. The representatives of the World Bank did everything possible to discredit the procedure created in London. Some representatives of developing countries tried to obtain the maximum in financial terms without assuming any serious commitment regarding the reduction or elimination of substances controlled by the Protocol. The Global Environment Facility (GEF) was created and located in the World Bank, and, in what appeared a suspicious move, included ozone depletion among its fields of competence. It seemed, at first, that the Protocol and the London amendments were headed towards failure.

In spite of these bad omens, the Executive Council started work in Montreal. A director of the Fund was appointed and disbursements began. Standards for project approval were established. The development of country programs was encouraged and the active participation of executing agencies (the World Bank, UNEP and United Nations Development Programme) was also promoted. The independence of the Fund with respect to UNEP was strengthened. The Executive Council, finally, assumed its full responsibility in the first months of its existence. Clear channels of collaboration between the developed and developing countries were established and the Executive Council made decisions by consensus with great efficiency. The energy and will of the Executive Council in the fulfillment of its mission allowed, little by little, the executing agencies to adapt their own procedures to the demands for prompt action that circumstances required. The design of the projects was simplified and their implementation accelerated.

The Fund of the Montreal Protocol has demonstrated that cooperation is indeed possible, that when the process of cooperation is real, it is not a question of donors and recipients but rather of countries, which, even at different degrees of development, share a common objective and decide each to contribute something to achieve such objective, be it financial resources, political will, technical or scientific knowledge, or any other element. In this sense, the protection or sustainable use of the environment enjoys today a propitious opportunity to help stimulate international cooperation where the political will of all and the generosity of those who have most can help make the Earth a clean, lovely and valuable planet for all of us, for our children, and for the children of our children.

THE MONTREAL PROTOCOL: WHOSE MODEL?

Ashok Khosla

The Montreal Protocol is certainly a successful example of international cooperation. My job here is not to pat ourselves on the back, but to look at the insides of that black box and see how much of that example might be applicable or serve as a model in other fields requiring such cooperation. I will tread a little bit into the territory of other papers which look at the negotiating process, but it seems to me that it is entirely of one piece when one is negotiating and implementing a new regime. We must look at the entire process with a fundamental understanding of what is needed to make such things work.

We also should be careful with terminology. When talking about things like international cooperation, we must not run away with the idea that there has been two-way reciprocity, or a sense of equality, between the parties concerned. This has not really existed in the past. There has not been so much cooperation as a one-way relationship in which the dominant partners get their way. "Industrial cooperation" and "development cooperation" are often simply euphemisms for aid, access to inputs and markets, euphemisms to get others to tow your line. This is not real cooperation. Rather, I will be talking about regimes that actually mean cooperation, not about such arrangements as the World Trade Organization (WTO), which I do not see as an example of cooperation. It is, rather, a way of making sure that those who have, those who are strong, negotiate the kinds of regimes that suit them regardless of the needs and opposition of those who do not have. We will talk about cooperation, cooperation that leads to an ultimate ideal regime in which everyone comes out a winner and everyone benefits.

In an inequitable world where there perceptions and expectations differ, it is quite easy to forget the views of others. What Dr. Tolba and Ambassador Mateos have described is an important example of how negotiating processes and regimes can be elaborated in order to let people get to know each other better and understand each other's needs. Unfortunately this is not always the case, but the Montreal Protocol can serve as an example to help us learn how to avoid the usual outcome in which the weak get the short end of the bargain, whether in terms of resources, political power or access to knowledge. Genuine cooperation is of mutual advantage, a win-win situation. This also means that a variety of outcomes should be possible. For instance, in mutual cooperation, it should be possible for one or more of the parties to opt out. In our present regimes that really is not possible. The imperatives under which our policymakers, both in the North and the South, must act make it very difficult to envision any actor actually saying "this is not for us and we wish to terminate our involvement." Cooperation also implies not only fair deals, but also elements of reciprocity, mechanisms for redressing imbalances as they occur, and freedom of choice. China, for example, has said as much in the past as have only a few other countries, but they have had to be vague and very strong in doing so. I will come back to this fundamental point later on in the discussion.

Thus, on the one hand, the Montreal Protocol is a very interesting and powerful mechanism that has shown us that it is possible to bring large numbers of diverse interests and viewpoints together and arrive at a valuable outcome that everyone sees is better than what existed before. But the Montreal Protocol is of limited value as a model for future regimes, and I will attempt to show what future regimes should look like, how they should be moulded and structured to make them look like the Montreal Protocol in order to ensure similar success.

The Montreal Protocol is of limited value because it had a very simple and fairly direct set of objectives. Future issues, like carbon emissions, climate change, biodiversity and others are much more complex. For instance, if I were to say that the Montreal Protocol is like the world deciding that smoking is bad and requiring it to be given up, applying this approach to the case of climate change would entail requiring people to give up a certain part of their food. Now, on the one side we are talking about substances that are not really essential to the well-being of humanity, to the survival of the planet and of life on it. It was therefore possible to come through with very narrowly conceived, and very focused agreements that were achievable. On the other hand, when I am talking about climate change or biodiversity, I am talking about giving up something that is essential to life itself, or giving up a part of it. This becomes more difficult because there are a variety of life support systems that have to be approached very differently. In a sense, we now must translate the kinds of things we have to do with respect to climate change, biodiversity, trade, the environment, and a diversity of other things into elements that we can tackle along the lines of the Montreal Protocol.

The Protocol was relatively successful because very direct factors underlay it. The ozone problem was relatively easy to negotiate because the Montreal Protocol was single-purposed, narrow, and well-defined; it had clear national constituencies and did not face much fundamental opposition: even the commercial interests were close to a series of technological breakthroughs which made it interesting for them to change. For competitive reasons, there were industries that may have been interested in going beyond chlorofluorocarbons (CFCs). There was, of course, a certain amount of inconvenience involved in the changeover, but it was not a major hurdle. The problem rested on very clear and well-defined scientific evidence and could rely on technical solutions rather than on a large number of political and social structural changes.

The Montreal Protocol dealt, in some measure, with symptoms and not causes. Of course, the CFCs were the cause of the problem, but clearly it meant that there were well-defined causal linkages through which you could quickly identify actions that would lead to results. It addressed issues that I would not consider as critical as many of the other regimes will have to address. It generated drama, there was a big ozone hole up there. This made it very easy for the public to climb on the bandwagon and feel that they were dealing with an immediate problem. It did not really require fundamental change in lifestyles. We will still have our refrigerators and air conditioners. It may cost us a little more but one will not have to give up a second Volvo or any of the other lifestyle changes required by climate change and biodiversity. It balanced, to a great degree, many of the benefits and costs of compliance and, to that extent, it made life a lot easier. Basically, the Montreal Protocol met the needs of the most powerful parties promoting it. It was in accordance with the wishes and the perceived self-interests of those who first pushed it as an agenda item on the international scene, and who then felt that it was their problem and that they held the solutions. For all these reasons, the Montreal Protocol was something that was doable. The climate change, biodiversity, conservation and other similar issues will not be so amenable.

One reason behind the failure of international negotiations is that we still live in a world in which those that have the capacity to act remain in a 1960s mind-set or, rather, mine-set: What's mine is mine and what's yours is yours. They are still thinking that they know all the right questions and have all the right answers. For these reasons, the West and the North have been able to impose on the rest of the world much of their conception of what constitutes the good life. In this way they have developed concepts like way-of-life, private sector, market, the intellectual property rights, the question of incremental costs, the area of jurisdictions, and the way in which you negotiate. For instance, Rio and the process that led to other conventions that followed Montreal were modeled after the Montreal Protocol in that they were all single-purposed. Although there were very deep linkages between the matter under negotiation in different regimes, like the connection between climate change and biodiversity or between climate change and the Montreal Protocol, these agreements were to be negotiated separately and distinctly because this made it possible to break up the process into a controllable and useable form, which, of course, suited the needs of those who wanted to control the outcome of the negotiations.

This system seldom allowed a complete southern viewpoint. What is a southern viewpoint? After all the South is a very complex set of entities. In the South there are a lot of Norths, countries which are represented internationally not by people who understand the needs and perceptions of our people, but by representatives who hold views, expectations, aspirations, understanding, and knowledge more closely related to that held by those with whom they are negotiating than by those whom they represent back home. This also raises the question of legitimacy. Whose interests are being represented? In the case of the Montreal Protocol, it is not very difficult to answer those questions because they applied to virtually everybody. Everybody gets skin cancer, cataracts and suffers from immune problems; everybody loses out when agriculture is destroyed, so representation was not really a problem. This is not true with the new regimes. We must be very careful not to get into a situation in which we think that we have the optimum solutions simply because one hundred-odd southern delegations are involved in the process. Those delegations normally consist of upper middle class individuals who represent only a minor portion of their countries. Furthermore, these people are often trying to get jobs in the World Bank or the U.N. or trying to get their children jobs in these places. I believe that we have to be careful when interpreting messages from previous experience when designing the future regimes.

One great thing that has come out of all these processes, starting with the Stockholm Conference in 1972 but gathering momentum since Rio, is the role of civil society. Scientists, engineers, the voluntary sectors, the technical sectors and the private sector were brought into these negotiations. The image that comes most clearly to my mind when I think about the global negotiating process, is a casino in which you have a game going on with many players. Those on the receiving end, usually the South, are not in a position to win or even to break even. It is a game in which the winner gets to change the rules as the game goes along, which is quite frightening. The Montreal Protocol, for example, has had many amendments. A few rested on clear, scientific evidence, but I believe many more were added for commercial reasons, for reasons that were extraneous to the survival of life support systems. These rules are like a moving target; and the South is not very good at hitting moving targets because it does not have the opportunity to train its guns regularly. The worst part of this game is not that you have the short end of the bargain, but that there is no way to get out of it because our political leaders are not in the position, or do not believe that they are in a position, to say "this is not our game". Ultimately, whatever comes out is at the mercy of the house which has written the rules. Before these regimes can be equitable, I believe that we need to learn a lot more than we will ever be able to from the Montreal Protocol.

So what does an ideal regime look like? It will have to be more systemic, more all-encompassing, with better links to all the problems we have caused through runaway science, technology, and global economics. None of these are bad in their own right, they brought great good for all of us, eliminating hunger and disease in many parts of the world. But there are still two billion people, maybe more, who have essentially been left out of all the progress that one calls science, technology and economics. These two billion constitute forty percent of the population of the planet. We cannot rest on our laurels until we realize that future regimes will have to encompass them and their interests. We need a regime that also allows the South, the downtrodden, the marginalized, to start writing the agendas and to say "These are our problems. How about righting them?" The South may appear to be always saying that we are poor, we need more money or more technology, but nobody has ever picked these issues up and made them part of the agenda. Just as some great scientists or constituencies are able to come to the U.N. or to the international community, present emerging problems and impart us to deal with them, some of the emerging problems should be arising out of the perceptions of other people too. We do not have the mechanisms to do that. Our mechanisms must be much more systemic and we must recognize the greater issue of interdependence if ever we are going to set different kinds of agendas that involve different actors, and negotiate these agendas with a greater sense of equality.

The Montreal Protocol was great at achieving this shift. It represented the archetypal example of interdependence. People from the North were creating all types of pollution which caused great changes in the South. That is the bottom line of all new regimes. We must also recognize that global change is not all that is in the cards. We can see that there will be great impact on the geopolitical future of our planet, and certainly a great amount of armed conflict, arising from the scarcity of resources. This will lead to problems of access to natural resources, diseases, and population migration. If certain countries were concerned by the problem of boat people in the 1980s and 1990s, then they have not seen anything yet. When sea level rises, hundreds of millions of people will be looking for new homes. We will see what the large scale impacts of our lifestyles has had on other people and the rebound effect of these impacts. I believe that we will experience these changes, if not in our lifetime, then very soon in the future.

The Montreal Protocol did not teach us that we need to view things on a larger scale. The urgency of viewing environmental problems in greater dimension must not be overlooked in future regimes and negotiations. The Montreal Protocol is an applicable model. It is a terrific model, and I do not mean to knock it in any way. But it is a limited model, and we must understand it fully in order to integrate its strengths, and rectify its weaknesses within the context of new regimes. I feel that the way to do this is to break up the new problems into ozone-like problems so that each component can then be dealt with. Ultimately, we need to improve negotiations by allowing people from all over the world to set agendas, by improving their negotiating ability and feelings of reciprocity, and by building the capacity of all people to identify their problems and find adequate solutions. Building that capacity has not been an adequate part of even the Montreal Protocol. It was a reactive protocol. It was not able to tackle the issues of building up scientific capabilities, of getting developing countries to develop the institutions, laboratories, frameworks, and workshops which can help understand and respond to developments.

A new refrigeration cycle is going to come from the South because the North has what it needs. New regimes dealing with responses to energy cycles and climate change must come from the South because there is no reason for the North to do so. They have reneged on all responsibility for it. It is, therefore, in the interest of the North that the South find and negotiate solutions, but the North does not seem to be able to understand that. As for integration of research and science into policy and technical diplomacy, I think we need to recognize that the job is not yet done. A lot of work is still needed towards building up capacities. The longer term vision we all need is that which considers that what is good for me also has to include what is good for you.

COMMENTS ON AMBASSADOR MATEOS' AND DR. KHOSLA'S REMARKS

Victor Buxton

I would like to start off my observations by saying that there are many perspectives one can take when talking about the Montreal Protocol. Dr. Khosla has one; many others have differing viewpoints. I happen to be one of them. I come at this from another perspective altogether. I shall look at these issues from the perspective of the Montreal Protocol as a new model for cooperation, and will start with a little walk in time while addressing some of the issues that Dr. Khosla has raised.

As Dr. Tolba has said, shortly before September 13 1987, we were not sure that there would be an agreement. I therefore take issue with the idea that this was an easy and simple protocol. It was ten years in the discussion phase and anyone that participated in the "Holy Wars" that led up to the Vienna Convention recognizes that this was no simple undertaking. A multi-billion dollar industry was at risk from a financial point of view, at least that was the perception at that time. We do no one a service by implying that this was an overly simple endeavor.

I also do not believe that the Montreal Protocol was reactive. If anything, it was proactive. There was no scientific consensus on this issue. It was really a question of risk management, the downside of which was too difficult to contemplate. The Vienna Convention, though a major step forward, was in many ways, a failure. It was like a bicycle with no pedals because we came to Vienna thinking we could do two things: either ban aerosols or cap capacities. We could not agree on either one. Yet, we went away with a great achievement: we agreed to cooperate and to work together on the next step. We pointed out that creating partnerships based on trust and understanding was a prerequisite to moving forward.

I too would like to make a few observations on definitions. First of all, the Montreal Protocol is a consensus process, but my definition of consensus might be slightly different than others. I do not believe that consensus means agreement.

Consensus means a willingness to go along with a majority viewpoint for the overall public good. Consensus means that you can be part of a consensus but still fundamentally disagree with what is there. You recognize that your disagreement stems from perhaps your own self interest and that there is an overall greater good that perhaps the collective wisdom is defining for you.

As for the partnership process, first of all, when we are talking about the relationship between developed and developing countries, we should also recognize that there is a developing country within every developed country. The same dynamic forces that happen within a global context happen within a national context on an ongoing basis and have to be addressed. The partnership of the Montreal Protocol was not easily achieved. For example, within the scientific community, the European Community scientists were convinced that this was all a hoax and that there was nothing there. The North American scientists were convinced that this was a real phenomenon and that it needed to be addressed. Huge bridge-building had to be undertaken during this process. Even economists disagreed with the rate of build-up of the chlorofluorocarbons (CFCs). Anyone present at the Rome meeting in the early 1980s, could witness vitriolics that did not bode well for any future agreement. There were differing viewpoints on technology and its possibilities, and whether or not substitutes for CFCs were theoretically possible. There were great debates on this issue in the early days. Trying to get producers to cooperate with each other on any kind of regime was a mammoth undertaking. The list goes on, even within and between internal governmental departments.

Turning to the intergovernmental forum, we had different blocks of countries all approaching the negotiations from different vantage points and for different reasons. The United States came forward with all kinds of scientific evidence and all kinds of turmoil from within its ranks. The European Community was all over the map with its members divided as often as they were consolidated, and often had to be brought together by political edict. Nordic countries were at the vanguard and pushing us, with the non-governmental organizations (NGOs) putting forth extreme positions. Canada, New Zealand and some of the others tried to play the role of honest broker. The Russians were somewhat of a wild card, not really knowing were they were at which time and on which issues. The Group of 77, the developing countries, were very interesting because they were coming to that process through no fault of anybody, but with a lot of difficulties. They did not have any inherent or indigenous scientific data. They had very limited ongoing production. They saw their development prospects threatened if they were unable to capture the chemicals that were on the table. All of this formed the dynamics on which the challenge was to develop a partnership process.

Now let us superimpose on that some of the other issues that were taxing the partnership process. There was the conspiracy issue. Many thought that the United States had substitutes and were trying to pass a ban in order to gain market shares with their product. Others thought that the issue was exports and that some countries were trying to steal market shares from others. There were trade fears. Some thought that this was simply a method to redefine the trade regime. Others thought that this never could happen unless we could arrive at an economically-levelled playing field, and this was impossible. All of these points serve to illustrate that the Montreal Protocol process and the partnerships that formed were really hard fought battles. I think that we have learned a lot and that the lesson will live on forever. We need to carry forward.

What did we learn about the partnership process? I think, and Dr. Tolba pointed this out in his remarks, it has to be incremental. We must go slowly and be willing to compromise. All have to participate in consensus-building along the definition that I gave you. There were ten years of dialogue previous to the Montreal Protocol. What did that achieve? I believe it led to trust and understanding, two prerequisites to partnership.

In London, in 1990, we increased the scope and stringency of the requirements instituted by the Protocol, but that was not the most important change. What we saw in 1990 were ethical revisions. People realized that if the Montreal Protocol and future regimes were to be successful, they needed to be predicated on mutual need. We were dealing with global issues that can only be sustained or remedied by global action. Unless everyone that could have a significant action as part of the process were implicated, no deal would be possible. That realization gave rise to an historical first: the Multilateral Fund, which deserves to be emulated many times in the future.

It is easy to look at the Protocol and say that it does not address all social inequities. However, I am not sure that is the right way to look at it. We have to look at it from another perspective, from the perspective that spaceships do not have any passengers, just crew-members; we are all in this together. Yes, some of us may have more or fewer advantages than others, but what is at stake is the well-being of the planet. This is absolutely critical for the model that we use to create agreement. I do not argue that climate change is not a tougher nut to crack, but the Montreal Protocol provides a lot of positive lessons on creating partnerships, on creating political will, on creating champions, on getting industry, governments and NGOs to work together. If countries continue to look at the Protocol from the vantage of "what's in it for me", then we have a problem. I do not agree that we have to look at this in a broader context, that we have to look at the social and economic issues on the same level as the environment. In a petri dish, the bacteria grow on a healthily nutrient-rich agar medium until they reach an equilibrium, then they will stop growing. If you think for a moment that perhaps the bacteria can be equated to population growth, and that the rate of population growth is a function of economics and well-being, you can very easily poison the agar slide and the system collapses in a non-retainable way. For many people, the hole in the ozone layer was Nature's veto. It showed that we had reached that point where we could not go much farther. We still have a lot to do and have to work together.

Ambassador Mateos made some correct observations about official development assistance and its implications in Rio and in Stockholm before it moved to 0.7 percent of gross domestic product (GDP) as a rate of contribution. Very few countries have done that. That's the bad news. The good news is that the Official Development Assistance (ODA) has changed very dramatically over the years. The level of conditionality has dropped dramatically. There is also a recognition, as Dr.

Khosla has mentioned, that the way is not snapshot fixes in terms of ODA but long-term institutional capacity-building. That requires a much larger time horizon; ten or more years of financial commitment as opposed to two or three years.

THE MONTREAL PROTOCOL: THE FIRST ADAPTIVE GLOBAL ENVIRONMENTAL REGIME?

Edward A. Parson

Does the Montreal Protocol represent a new model for the negotiation and operation of international regimes? Other papers have identified several innovative aspects of the Protocol, of which I will concentrate on one. I propose that one fundamental respect in which the Protocol is a new model for international environmental diplomacy is that it is the world's first *adaptive* global environmental regime. This adaptive character is related to the "dynamic and flexible" character of the Protocol that several other participants, including Dr. Tolba and Ambassador Benedick, have identified, but poses more specific conditions. An adaptive regime is one that, in pursuit of an unchanging goal, does two things. It supports identification, synthesis, and assimilation of changes in relevant knowledge; and it incorporates the results of changed knowledge into revisions of control measures, policies, and institutional arrangements. Put another way, in articulating its original goal, an adaptive regime incorporates the insight that what is needed to attain the goal cannot be fully known at the outset, but must be progressively adjusted over time.

What makes the Montreal Protocol adaptive lies partly in the text of the treaty and partly in practices that have developed since 1987. In the treaty, Article 6 specifies that at least every four years, the Parties must assess the control measures on the basis of available scientific, environmental, technical, and economic information, and that at least one year before each such assessment, the Parties must convene appropriate panels of experts in each of these fields to report to them. These requirements, and the delicate balance of responsibilities and communication between the Parties and their assessment panels that has developed, have been the principal engine driving the progressive strengthening of the Protocol since 1987.

The enactment of these measures represented commendable foresight and initiative of those who worked on the Protocol in 1986 and 1987, but I believe they also reflect some measure of historical accident and good luck.

Good luck was involved in two ways. First, it is important to note, as others have alluded in their contribution to this symposium, that the 1987 control measures— principally a commitment to cut production and consumption of chlorofluorocarbons (CFCs) by fifty percent—were negotiated at a time when, on the one hand, the urgency of doing something to protect the ozone layer was widely recognized but, on the other hand, the stringency of measures necessary to protect it was not known. The authoritative international scientific statement at the time was the 1986 report "Atmospheric Ozone", widely known as the "Blue Books", which was sponsored by several agencies including both the World Meteorological Organization (WMO) and the United Nations Environment Programme (UNEP) but principally initiated and funded by the National Aeronautics and Space Administration (NASA). This authoritative three-volume summary of the scientific knowledge of the time was more survey than policy assessment, but its projections of the consequences of various scenarios of CFC use were widely cited as the best policy-relevant knowledge of the time. These projections were widely summarized into two simple messages. On the one hand, roughly constant CFC emission levels would result in only small depletions in total global ozone, of the order of a few percent, and these losses would be even smaller if other anthropogenic emissions, such as carbon dioxide and methane, continued to increase. On the other hand, substantial continued growth in CFC emissions would bring large losses of global ozone, of the order of ten to twenty percent. This report was in its final stages of preparation when the British Antarctic Survey reported their observations of the Antarctic ozone hole from Halley Bay. In a brief paragraph added in final editing, the report merely noted these shocking observations, and said that their cause or significance could not yet be assessed. Although the scientific expedition that yielded the decisive observations attributing the hole to CFCs was underway even as the negotiators met in Montreal, the question of the cause of the hole was unresolved as the Protocol was signed.

Consequently, if I might presume to make a retrospective rationalization of how the 1987 negotiators came to the fifty percent reduction, it might be as follows. If CFCs are *not* the cause of the Antarctic ozone hole, then the appropriate control measures look something like a freeze to prevent growth above present levels, as had been adopted as negotiating positions by both the European governments and the American chemical industry, based on their respective interpretations of the Atmospheric Ozone report. But if CFCs *are* the cause of the Antarctic ozone hole, then the appropriate measures look like very stringent cuts, perhaps complete elimination of the chemicals, as had been proposed by the United States and the Toronto Group. Similarly crucial and unresolved questions existed on matters of the technical feasibility of reducing CFC use, as anyone who recalls the feeble technology exhibits in the foyer of the meeting center here in Montreal ten years ago will realize. Under these conditions, it does not appear far-fetched to describe the agreement of 1987 as an agreement to split the difference (between holding to present levels and total elimination), and to revisit the question in a few years with continued scientific and technical input. This agreement, forced by the provisional character of the measures undertaken in 1987, was what left us with such an effective process of periodic review and assessment.

The second respect in which good historical luck was involved in shaping the current situation concerns the form of the assessment process. The most salient example of an international assessment process in the minds of the negotiators was the Blue Books, plus the subsequent contribution of leaders and participants in that process to informing the negotiations. This contribution was viewed as so valuable by all Parties that the negotiators agreed, apparently with minimal discussion, to replicate the process of the Blue Books in the assessment panels established under the Protocol. For the Atmospheric Science panel, this decision amounted to a straightforward application of a proven model to a very similar task, albeit with some required adjustments of mandate and participation. But in the other domains, particularly in technology and economics, the Blue Books model was generalized to a very different domain of questions, resulting in the establishment of an unprecedented body, the Technology and Economic Assessment Panel (TEAP). It is a powerful indication of the esteem in which the 1987 delegates held the Blue Books that they established a panel in its image to address questions of technology and economics, where the line between technical and political argument is much harder to draw defensibly, and that they did not try to assert direct control.

These decisions, through whatever combination of intelligence, foresight, and luck, have left us with an assessment process under the Protocol of unprecedented effectiveness.

Several features contribute to their striking success. First, all panels operate in the basic spirit of separating assessment from management, while still providing assessment outputs that are policy-relevant. While this is a Zen Koan, never fully realizable, the panels have been impressive in their ability to skate along the border and avoid significant political controversy over the substance of their reports or the process of their work. The success of TEAP in this regard, dealing with questions for which the separation of assessment from politics is much more difficult than for atmospheric science, is particularly impressive. Second, that panel members serve as individuals, in their scientific and technical capacities, *not* as representatives of a Party or other constituency, has contributed both to the high scientific and technical standards they have achieved and to their effective independence from the political differences of view that have at times divided the Parties. This independence has been further enhanced, particularly on the TEAP where it would be most difficult to achieve, through process rules requiring collective deliberations with arguments conducted only on technical or scientific bases, anonymity in reports, and the prohibition of bound votes.

Perhaps most important in sustaining the effectiveness of the panels has been the skill exhibited in defining the scope of questions they undertake. By and large, for the atmospheric science panels, the policy-relevant questions in which they summarize the results of their works for policy-makers have been "if-then" questions, in which the "if" denotes specific measures the Parties might decide to undertake, such as specified schedules for further restrictions of ozone-depleting substances. The "then" expresses consequences of these hypothetical decisions in terms of an environmental measure that is sufficiently simple and stable (i.e., numerical estimates not to fluctuate much from year to year), and that is widely accepted to be important. The most often-used measure of environmental consequence so employed has been the projected future time-path of total stratospheric chlorine. The acceptance of this measure as sufficiently policy-relevant has enabled assessments to be agreed among scientists with enough consensus and authority that the charges of political influence that arise occasionally in the climate-change assessment process have not arisen in the ozone science assessment panel. Defining the questions to address is even more delicate for the Technology and Economics panel. They have emphasized judgments on the level of substitution or reduction that is technically and economically feasible in specific use sectors by specific dates, and where the terms of reference, definitions, and criteria—particularly when these are likely to be contentious—are specified in advance by the Parties based on informal consultation with panel members.

It is important to note that no assessment process, however effective, can or should eliminate disagreement or controversy over the appropriate policy course. The ozone panels have not done this, as current controversies over the treatment of hydrochlorofluorocarbons (HCFCs) and methyl bromide amply reveal. They have, though, removed certain bounded scientific and technical questions from the domain of political controversy to a separate forum where they can be resolved, or partly resolved, on scientific or technical grounds. Disagreement and controversy are not eliminated, but the scope of policy positions for which putative scientific or technical justification can be credibly advanced is narrowed. The remaining zone of disagreement is more purely political, and less confused by the confounding of differences of scientific or technical judgment with questions of political preferences or values. In addition, the TEAP has in some cases made a distinct and novel form of contribution to the regime. It has provided a vehicle to promulgate and evangelize relevant technical innovations direct to the affected Parties, speeding the dissemination of knowledge and of innovations rapidly among specific sectors worldwide. This first-order effect supporting innovation has a second-order effect on the political process, as firms that might have opposed the process through their domestic governments are turned into supporters by the realization that acceptable, or even desirable, alternatives exist to their present way of doing business.

For the general development of adaptive regimes for global environmental issues, I would propose that the first ten years' experience of the Montreal Protocol offers four simple lessons.

1. The Flywheel

If an environmental regime is pursuing real action on a serious problem—that is, if it seeks more than either symbolic action or the international ratification of what is already nearly universally agreed—then every participating government, however keen its commitment to the global environment, will experience times and situations when it becomes a less enthusiastic or even obstructive participant. While uneven enthusiasm is to be expected of any participating nation, the experience of the United States in the ozone regime illustrates it particularly vividly. Last-minute reactions within the federal government threatened to derail U.S. support for the ozone regime three times: at the 1985 signing of the Vienna Convention, during final negotiations of the 1987 Protocol, and at a late stage in negotiations of the crucial 1990 amendments.

Given this inevitable inconstancy of all national participants, a continuing international process is essential. If the international process has a regular schedule that cannot be easily or arbitrarily delayed, and a high enough profile to embarrass Ministers, then it can develop enough inertia to oppose periodic lapses of will among major participating nations, and provide a vehicle for different governments to compete over time for international leadership. Some have called this aspect of the Protocol a *ratchet*, but I would argue that the *flywheel* is a more apt image. It is not the case, nor should it be, that an international regime can never reverse direction; sometimes advancing knowledge may indicate that earlier enacted measures were misconceived. But the process should build momentum that does not always depend on continued pushes from the same participants, that can smooth variations in individual inputs and resist short-term lapses of domestic political will. Institutions with stable mandates and funding, regular schedules of meetings and administrative requirements, and the engagement of multiple sectors and bodies in linked systems of deliberation and decision-making, all contribute to this smoothing function.

2. Seek And Ye Shall Find

This lesson poses a paradox of adaptive management. It is based on the observation that advance estimates of the cost and difficulty of making a technical change, imposing a regulation, changing a process, or reducing a substance or activity, are extremely unreliable. In part, this unreliability reflects the fact that when a present product, process, or technology is working and profitable, it is not worth looking hard for new ways of doing things unless under the threat of compulsion. In part, it reflects that advance cost estimates depend primarily on the expertise of those engaged in the present activity, whose interests may lead them to make very cautious or high estimates of the cost or difficulty of changing.

For these reasons, the cost and difficulty of meeting new environmental targets often turn out to be substantially smaller than were predicted in advance. Current research suggests that this is the case for many specific areas of the phase-out of ozone-depleting substances under the Protocol. But this beneficent result cannot be relied upon. Sometimes, when one starts looking at how to do something, one discovers that the problems are harder or more numerous than expected and costs turn out to be as high as, or higher than, projected. And sometimes, the world just looks different after making a change—new products or processes differ from the old ones in many ways, and it is difficult even to define retrospectively what the cost of meeting the environmental target was.

This unavoidable uncertainty means that environmental policy-makers must make regulatory decisions with very weak knowledge of how much they will cost. This uncertainty, which clearly prevailed in 1987, can introduce various paradoxes into the dynamics of interactions between policy-makers and firms affected by regulations. These are best illustrated by the attitude of a manufacturer of an important piece of CFC-using equipment. In anticipation of the CFC phase-outs, his firm had developed a major innovation that eliminated CFCs, improved performance, was cheaper to operate, and consequently was a hugely profitable market success. His assessment of this experience consisted of a lengthy period of enthusiastic boasting about the innovation and the profits and leadership position it brought to his firm, followed by the exhortation that on all accounts the Protocol negotiators should not put them in a position where they would have to do it again.

3. The First Ten Percent Is Easy, The Last Ten Percent Is Hard (or Different)

The process described above, of finding cheaper and better ways to innovate out of an environmental problem once you start looking, appears to generalize across single, or closely linked, technologies, processes, and firms. Changed thinking and practice, as well as specific innovations, can spread among firms or individuals who collaborate closely or who compete directly.

But solving environmental problems with diverse causes, such as ozone depletion or climate change, requires moving progressively to new chemicals, emissions, activities, industries, and technologies. Even when environmental policies are implemented through market-based measures that do allow decentralized shifting of effort among the activities under their scope, it is rare that the entire set of relevant activities are brought under treaty or regulation from the outset. Consequently, continued management of the problem normally requires progressively extending controls to new gases, processes, activities, or technologies. And each new area poses new technical, economic, and political problems. The new targets coming under the scope of an expanded regulation have not normally participated in the socialization that has brought changed thinking and practice to those firms and industries who are already in. Moreover, since prudent regulators tackle the easiest parts of a problem first, the technical and political problems posed by the new sectors will not just be as hard as the original sectors were, but harder. Methyl bromide is the case in point.

Indeed, it is plausible that the optimal strategy for any industry threatened by regulation that may be difficult or require changing practices, habits, and ways of thinking about its business, consists of two distinct stages. In the first stage, you dig in your heels and resist with all the force you can command until you decide that the regulation will inevitably be enacted, sooner or later. In the second stage, after your expectation has so changed, you reverse stance and compete to lead the pack in the new way of doing things. This reversal may even have the character of a conversion experience.

This process holds two lessons for those crafting environmental treaties and regulations. First, one should not expect that once one sub-domain of a problem has come under management, all subsequent difficulties will go away. Any incremental approach will bring new participants and new potential opponents at each step. Not having experienced the changes of view, or the conversions, of those assimilated at previous steps, the newcomers can be expected to fight just as hard as their predecessors did. Second, a regime moving into a new area should do it in a way that avoids unnecessarily inflaming potential opponents and that maximizes the benefits available from the lesson above, namely that one is likely to find better ways once one has started looking. One element of this approach involves setting relatively near-term interim targets whose probability of being achievable is high, whatever the ultimate goal is. These interim targets should be demanding enough to get people's attention, to re-direct development effort and to set in motion the forces that so often lower the cost of change. But they should be easy enough that they only rarely turn out to be unattainable and must be reversed. Still, if the regime really is managing adaptively, sometimes reversal will be necessary. The required paradox is to maintain a commitment sufficient to force real effort from the regulatory targets, while being able to back off on the infrequent occasions where it is necessary.

4. Do Not Demand Perfection

Finally, in a regime that is pursuing real policy change, with real difficulties and obstacles, there will be occasional lapses, including failures of compliance and targets that must be changed. An adaptive regime must be able to tolerate less than universal compliance without unraveling. A regime that remains functional only with perfect compliance will either break apart or come progressively to be re-defined so that compliance is meaningless. The requirement is to maintain commitment sufficient to force real effort, without being so rigid that a single instance of failure brings the regime down.

In conclusion, the Montreal Protocol will only truly be a model of a new way of negotiation and sustaining global environmental regimes if it is imitated. Thus far, regrettably, it appears that the Protocol is more honored with words than with deeds. While the Conventions on climate and biodiversity copy many aspects of the Protocol, in the aspects I have discussed here as being essential to its character as an adaptive regime, both are much weaker. Perhaps there is a paradox of innovation in regimes deeper than those I have discussed here, that may lead any effective innovation to have a limited viable life. Perhaps climate and biodiversity failed to adopt the most effective innovations of Montreal because Parties who wished to limit the effectiveness of the regimes observed how effective these innovations were for the Protocol. It would certainly be unfortunate if such motives have influenced the design of review and assessment processes for the climate and biodiversity conventions, and indeed unnecessary. The record of the Protocol shows that while the panels have been independent and influential, they have not usurped the political authority of the Parties to decide as they choose. The Parties have at times declined to take steps that many argued were clearly, though of course implicitly, favored by the panel reports. Still, innovators may need to continue generating new ways to move regimes forward, against the procedurally powerful opposition of those who do not want it to happen.

COMMENTS: THE ATMOSPHERE AS GLOBAL COMMONS

Marvin S. Soroos

Today, you have heard from numerous U.N. officials, diplomats, and government advisors. I come before you as an academic, an educator, and a writer. As a social scientist, I have been specializing for twenty years on the subject of international environmental law and policy. I have been especially interested in matters pertaining to the management of global commons, such as the oceans and seabed, outer space, and Antarctica. The atmosphere can also be described as a global commons, although it is perhaps not as clear-cut an example as the others. Use of the other global commons is to varying degrees governed by major framework treaties, namely the Convention on the Law of the Sea of 1982, the Outer Space Treaty of 1967, and the Antarctic Treaty of 1959. These treaties have been the foundations for the international regimes that evolved to address problems related to the use of the commons. There is no similar foundation agreement that referees use of the atmosphere. In fact, the atmosphere has hardly been viewed as a legitimate object of international environmental law and policy. Rather, distinct regimes have been created to address atmospheric problems related to several types of human-produced pollutants.

In my recently published book *The Endangered Atmosphere: Preserving a Global Commons*,[1] I analyzed the international responses to four atmospheric problems. Several speakers have compared the international regimes that address two of these problems—depletion of the ozone layer and climate change. I believe it is useful to compare these regimes to two others—the aboveground testing of nuclear weapons and the transboundary flow of acid-forming pollutants.

The first of these atmospheric problems that the international community addressed was the testing of nuclear weapons. More than five hundred aboveground tests were conducted between 1945 and 1980, with the United States and Soviet Union accounting for 215 and 219 explosions, respectively. The United Kingdom,

France, and China set off much smaller numbers of nuclear explosions. The cornerstone of the regime that was created is the Limited Test-Ban Treaty of 1963, which prohibited nuclear explosions in the atmosphere, outer space, and the oceans. The only two countries that continued conducting atmospheric tests were France (until 1974) and China (until 1980). No aboveground nuclear tests have been reported or detected since 1980. Thus, the ban on the testing of nuclear weapons in the atmosphere is widely considered to have evolved into a principle of international customary law. As a response to an environmental problem, the atmospheric testing regime is an open-and-closed success story. It has been less successful as an arms control agreement, as widespread underground testing made it possible to continue development of nuclear weapons arsenals.

The second atmospheric problem to be addressed is acidification due to what is called long-range transboundary air pollution (LRTAP). Such pollution is emitted into the air in one country and deposited outside its borders, either in other countries or in ocean areas. Transboundary air pollution has traditionally been dealt with on a regional scale, primarily in Europe and North America. The series of international agreements that address this problem, which were negotiated under the auspices of the United Nations Economic Commission for Europe, were precedent-setting and seem to have had an impact on negotiations on how to address the problem of ozone depletion. It is interesting to look at the development of this regime since the adoption of the Convention on Long-Range Transboundary Air Pollution in 1979, which is a typical, weakly worded framework treaty that lays the foundation for the later negotiations on binding international limits on air pollution. The 1985 protocol on sulfur emissions is notable for pioneering across-the-board percentage reductions in air pollution that are binding on all parties. The Revised Sulfur Protocol is significant for being based on the concept of "critical loads" for the deposition of pollutants, which were used to set differential targets for the nations involved. Despite all of these agreements, the LRTAP regime has been only partially successful in mitigating the problem of acidification in Europe.

The regime that addresses the problem of ozone depletion that we are talking about today is perhaps the most auspicious success story in international environmental diplomacy. It is a case in which governments heeded the warnings of scientists about the emergence of a potentially catastrophic environmental threat by agreeing to phase out most of the chemicals that were believed to be causing it. Progressively stronger international regulations of these substances were adopted even before the harmful effects of increased dosages of ultraviolet radiation on the environment and human health were readily manifest. There is no need for me to elaborate on the specific provisions of the agreements that have been struck, since most of you who are gathered here are intimately familiar with the Montreal Protocol and the series of amendments and adjustments that were adopted later. Suffice it to say, if there is general compliance with these agreements and additional measures are adopted to phase out the remaining ozone threatening chemicals, it appears that the ozone concentrations in the stratosphere will be back to natural levels by the second half of the next century.

The international response to climate change is the last of the four atmospheric regimes that I examined in the book. Even though climate change is arguably the most serious environmental threat that confronts humanity, the international community has achieved very little towards mitigating the problem. All we have thus far is a foundation agreement, namely the Framework Convention on Climate Change, which was adopted at the Earth Summit in 1992. Nothing has been done yet to regulate the flow of greenhouse gases into the atmosphere from human activities, such as the burning of fossil fuels and extensive clearing of forests. We can ask whether the gathering of the parties in Kyoto later this fall will be to climate change regime what the Montreal meeting ten years ago has been to the ozone depletion regime. I share my colleague Ted Parsons' hesitation about predicting such an outcome. I am not very optimistic that an agreement will be concluded in Kyoto which requires the parties to significantly reduce their emissions of greenhouse gases.

My pessimism about the prospects for a strong international climate change regime takes into account two factors that I do not believe others have brought up at today's colloquium. The first one has to do with domestic politics and the extent to which atmospheric issues have become politicized. Here I take an admittedly American-centric perspective, which I rationalize for the following reasons. My country played a strong role in advocating international controls on ozone-depleting substances up through the Montreal meeting in 1987. By contrast, the United States has been the leading opponent of any agreement that would establish a binding timetable for limiting the pollutants responsible for climate change. Curiously, while the Europeans were initially skeptical of the need to phase out chlorofluorocarbons (CFCs), they have been strongly in favor of taking decisive steps to address the threat of climate change.

The international response to the ozone depletion problem never became politicized in the United States in the way that the climate change issue has over the past year. Congress was strongly supportive of the negotiations leading up to the Montreal Protocol. By contrast, it is discouraging that the American Senate recently adopted a nonbinding resolution by a vote of 95-0 which advises the Clinton Administration against making any commitments at Kyoto to reduce greenhouse gas emissions that might harm the economy and do not also apply to developing countries. The latter provision, which is politically expedient because developing countries are not a constituency within the American political system, fails to acknowledge that the industrial countries are responsible for the lion's share of human additions to atmospheric concentrations of greenhouse gases.

I am also concerned about the efforts of a coalition of large industry and consumer associations to turn the American public against any significant agreement that might be reached at Kyoto. They have sponsored a rather small, but outspoken and highly visible, group of scientists who have used the print and broadcast media effectively to cast doubt on the seriousness of the threat of climate change, as documented by groups such as the Intergovernmental Panel on Climate Change. They appear to have been quite successful in persuading the American public that a decisive response to global warming can be postponed until further scientific issues have been resolved. I think that most of you would agree that this is a gross misrepresentation of the state of scientific knowledge on climate change. It is, however, a political reality in the United States and one that was brought home to me recently when I was asked to be a guest on a local talk-radio program.

The second observation is a more theoretical one. I would propose that there are four strategies that can be adopted to try to deal with global environmental problems. One of them is to *prevent* or *limit* the extent to which the problem develops, which is the category in which I would place the Montreal Protocol. This approach entails negotiations among many countries in order to agree on limitations on the activities that contribute to the problem. The remaining possibilities would allow the environmental threat to develop, but reduce societal vulnerability to the impacts. Thus, a second strategy is to *avoid* the harmful consequences of an environmental change. Recall, for example, the suggestion of Donald Hodel, a Reagan administration official, that people stay out of the sun to avoid the harmful health effects of increased exposure to UV radiation, such as skin cancer. A third possible reaction is to *defend* against undesirable consequences, for example by wearing broad-brimmed hats and sunscreens as protection against UV rays. Finally, there is the option of trying to *adapt* to whatever consequences materialize, such as by seeking medical treatment for the health effects as they appear.

I believe that one of the principal reasons the Montreal Protocol was successful was the widespread perception that prevention was the only realistic way to respond to the threat of ozone depletion. This is not to imply that the many other reasons mentioned today for the success of the Montreal negotiations were not also important factors, but simply to suggest the nature of the problem was favorable to a preventive solution. In contrast to the ozone depletion problem that would affect all nations in a similar way, but not necessarily to an equal degree, climate change will impact on countries in quite different ways. Small island states anticipate catastrophic consequences from rising sea levels and storm surges. In the United States there is a widespread perception that the impacts of climate change may be both positive and negative, and that a major effort to limit the amount of change can be delayed until the nature of the consequences becomes more apparent. Some economists maintain that it might prove to be less costly to avoid, defend against, or adapt to climate change, than to prevent or significantly limit it.

Investments in preventive strategies designed to minimize environmental threats contribute to the creation or maintenance of the common good. In the abstract this appears to be a highly prudent and rational strategy. However, the political context in many countries, including the United States, makes prevention, which benefits humanity as a whole, more difficult to sell to their citizenry than avoidance, defense, and adaptation, which work to the advantage of one's own country exclusively. The benefits from limiting emissions of greenhouse gases are shared with all states, while those derived from building sea walls to protect against rising ocean levels are enjoyed only by the country that invests in them. Thus, if adaptive responses appear to be possible, countries may be tempted to opt for them.

These are a few of the reasons why I am not optimistic about the prospects for a significant climate change agreement as the date of the Kyoto meeting approaches. It will be interesting to see how this issue plays out in the United States, the world's largest emitter of greenhouse gases. The Clinton administration will try to inform the American public about the seriousness of the situation through a White House conference to be held in October 1997. But it appears that when push comes to shove, the Administration is very cautious about getting ahead of what public opinion appears to be on this type of issue. If the Clinton Administration perceives a lack of support for a decisive response to the climate change problem in Congress and the broader public, I believe the United States will come to Kyoto with rather bland proposals. I do not wish to leave you with the impression that the successes of the Montreal Protocol were not important. However, I do not think that we should assume that it is possible to move from one regime to another and apply the same formula for success. Each problem poses a unique set of challenges that will require a distinctive international response.

The question arises whether there should be a comprehensive regime for the atmosphere that would be comparable to those that exist for other global commons, namely the oceans, outer space and Antarctica. In theory, perhaps yes. But the experience with negotiating a comprehensive law of the sea suggests that forging a similar foundation agreement for the atmosphere would be a protracted process that could span decades. I would prefer that the limited diplomatic resources of the international community be devoted to strengthening the existing problem-specific regimes, in particular to stemming the transborder flow of acid forming pollutants and to limiting the flow of greenhouse gases into the atmosphere. Furthermore, some additional work is needed to tighten the ozone depletion regime to phase out the remaining ozone threatening chemicals, in particular methyl bromide and hydrochlorofluorocarbons (HCFCs). Addressing the four atmospheric problems individually has proven to be a reasonably successful approach to managing the use of the atmosphere.

Notes

[1] Columbia, SC: University of South Carolina Press, 1997.

PART 4

THE ROLE OF TECHNOLOGY

TECHNOLOGY ASSESSMENT FOR THE MONTREAL PROTOCOL

Suely Machado Carvalho

The existence of the TEAP is a result of the visionary thinking of the original drafters of the Montreal Protocol. In 1987, the Parties to the Protocol recognized that over time, our scientific understanding of ozone layer depletion and its effects would improve. While the original control measures seemed both supportable by the scientific understanding of the day and a real challenge for industry, the architects of the Protocol realized they could turn out to be inadequate to protect the ozone layer, and much easier to achieve than industry believed or acknowledged at the time. They also realized that technological innovation would likely make future controls more technically and economically feasible, and that improved scientific understanding of the mechanisms of ozone depletion could demonstrate that more stringent controls could be necessary to protect the ozone layer. As a result, a process of continual assessment of science, technology and economics was built into the Montreal Protocol.

At the Meeting of the Parties in 1989, a decision was taken to create four Assessment Panels to advise the Parties on the changing scientific, technical and economic understanding of the ozone layer depletion issue. These were the Science Assessment Panel, the Effects Panel, the Technology Assessment Panel, and the Economic Assessment Panel. The Science Assessment Panel was charged with periodically reporting changes in scientific understanding of the condition of the ozone layer and the mechanisms of its depletion. The Effects Panel was responsible for keeping the Parties advised of improved understanding of the environmental effects associated with ozone layer depletion. The Technology Assessment Panel reported on new technologies being developed and commercialized to replace ozone-depleting substances (ODS) within the industrial sectors which relied on them for their businesses. Finally, the Economics Assessment Panel studied the economic consequences of ozone depletion and the economic viability of the alternative technologies reported by the Technology Assessment Panel. Because the bulk of Economics Assessment Panel work focused on the economic viability of alternative technologies, and because this issue was key to rapid implementation of alternative technologies, the Parties merged the Technology Assessment Panel and the Economics Assessment Panel in 1990 to form the TEAP.

What is the TEAP?

Today, over 300 experts serve on the TEAP and its TOCs and subsidiary bodies. Since its creation, over 600 experts from forty-five countries have participated in the assessment process. While members of these bodies have come from government and academia, it is interesting to note that most of the experts have come from the very industries which have been most directly impacted by the controls on ODS.

The Technology and Economic Assessment Panel, known as the TEAP, today consists of the following seven Technical Options Committees, or TOCs:
1) Aerosol, Sterilants, Miscellaneous Uses Technical Options Committee;
2) Foams Technical Options Committee;
3) Halons Technical Options Committee;
4) Solvents, Coatings and Adhesives Technical Options Committee;
5) Refrigeration Technical Options Committee;
6) Methyl Bromide Technical Options Committee;
7) Economic Options Committee.

In addition to the TOCs, the TEAP may establish occasional Task Forces and other Subsidiary Bodies. TEAP establishes these temporary bodies in order to obtain specialized expertise it may require to respond to specific requests from the Parties.

Technical integrity: how does it work?

At first glance one may question how affected industries could be objective. In practice, this arrangement has actually produced more rapid development and adoption of alternative technologies than would likely have been achieved if TEAP consisted only of government and academia. This arrangement resulted in industry being a key part of the solution to ozone layer protection.

Prospective members are nominated as individual experts by their governments to the Montreal Protocol Secretariat who forwards curriculum vitae to the TEAP for review and consideration. TEAP keeps the Secretariat advised of needs for experts based on the mix of expertise required to respond to requests from the Parties, and on the need to maintain balanced global representation. While this is often referred to as maintaining geopolitical or geographical balance, in reality a technology solution appropriate for one region may not be feasible elsewhere. This balance enables the TEAP to assess better the extent to which alternatives are technically and economically feasible throughout the world.

TEAP and all Subsidiary Body members work as volunteers under instructions from the Parties only. They represent only themselves, function as independent technical experts, and are expected to exercise objective professional judgement. Members are under strict rules not to accept instructions from employers, including government agencies. As a practical matter, members tend to be highly motivated and environmentally concerned individuals. A mix of expertise is maintained in order give the TEAP the capability to present objectively the full range of views on technical and economic issues related to the ODS phase-out. While the TOCs operate under the TEAP, all TOC reports are published without review or approval of the TEAP, and information from the TOCs is provided unfiltered to the Parties.

TEAP has been called upon to find solutions to some extremely complex situations. Assembling a multidisciplinary international team of experts to conduct a thorough assessment of complex technical issues is not an easy task. The following characteristics have been essential to TEAPs success:

1) teamwork;
2) mutual respect;
3) motivation; and
4) professional pride in being members of the first global consultancy community.

Why does it work?

TEAP has been successful for a number of reasons. As I mentioned earlier, most of the TEAP, TOC and Subsidiary Body members come from affected industries. One reason industry has been such a constructive partner in finding solutions to ozone depletion may have to do with an overall increase in corporate responsibility towards the environment, coupled with corporate understanding that innovation leads to increased competitiveness and new business opportunities. As science discovers that past practices which were believed harmless are in fact harmful to the environment, corporations are increasingly directing their innovative efforts towards improving environmental performance. Recognition that ozone layer depletion represents a true threat to the environment which requires real solutions has prompted a critical mass of corporate leaders to work together to find alternatives to their ODS uses, and to share them with others. These corporations offered their experts as TEAP, TOC and Subsidiary Body members. This contribution of employees' time and sponsorship of travel expenses has been crucial to the success of TEAP. It is this combination of technical independence, access to the world's foremost experts in industrial ODS applications, and global representation that has made the TEAP a powerful and invaluable resource to the Parties. They have come to rely on TEAP's highly educated and diverse membership, global perspective, independence, and technical integrity for advice. As a result, the Parties have confidence in TEAP's competence and objectivity.

Welcome results

While the ozone layer is far from repaired, the Montreal Protocol has been an unqualified success. The increasingly stringent controls on ODS production and consumption have been driven by science, and enabled by technology. As this connection between science, technology and public policy grows stronger, TEAP fulfils a crucial role in a process of international negotiation. The TEAP's work has brought very welcomed results. As a result of its efforts, mature and superior technologies have been identified, developed, commercialized, and successfully transferred between developed and developing countries. TEAP has also stimulated industrial innovation, and has appropriately recognised industry leadership and national efforts towards an ODS phase-out.

As a direct result of this process, TEAP has created a pool of experts in developed and developing countries. These world class experts are influential ambassadors in the professional networks, associations and societies of their respective countries and worldwide. The TEAP process organises, engages, cultivates, trains, supports, appreciates, respects, and rewards its members and former members. This creates a worldwide cadre of experts who inspire confidence in alternative technologies, and help drive implementation of new technology worldwide.

Looking toward the future

TEAP's work is far from complete. Important uses of ODS remain, and the deadline for the first controls for developing countries is rapidly approaching. To address these challenges, TEAP's future focus is directed towards:
1) increasing the pace of technology transfer and innovation;
2) replenishment of the Multilateral Fund;
3) the implementation process in developing countries.

Among the lessons learned are that simple solutions should be pursued first, and the hardest problems attacked last. The best solution should be selected from the options available, and specific solutions to specific problems produce good results. As a general rule, there are no "silver bullet" solutions and no universal drop in alternatives. Each ODS application must be evaluated in context, and solutions tailored to unique conditions.

The TEAP process has been crucial to the success of the Montreal Protocol. Because it represents a framework for independent, international technical collaboration, not only is it transferable to other Conventions but it is hard to imagine an international convention achieving the degree of success the Montreal Protocol has enjoyed without the existence of such a process. The Climate Convention is a prime example of a situation which would benefit from a TEAP-like process. Any science driven convention which depends on technological breakthroughs to achieve public policy objectives needs a TEAP-like process. The extent to which the TEAP model can be directly embedded in other structures must be evaluated on a case by case basis. However, probably the single most important characteristic of TEAP is that its focus is "Technology not Politics". While it may be difficult for politicians to rely on the technical assessments of independent experts, this focus is essential for the rapid identification and ultimate implementation of workable solutions. These issues and other lessons from TEAP should be carefully considered by those responsible for the success of other international environmental conventions.

TEAP TERMS OF REFERENCE

Robert Van Slooten

This paper sets out basic facts about the United Nations Environment Programme (UNEP) Technology and Economic Assessment Panel (TEAP). The objective of this paper is to answer the following questions:
1) What is the TEAP?
2) What are the Terms of Reference?
3) What are the criteria and procedures for selecting TEAP and TOC members?
4) How does TEAP go about fulfilling its Terms of Reference?

TEAP AND ITS TERMS OF REFERENCE

The members of TEAP are the co-Chairs of the seven Technical Options Committees (including the Economic Options Committee) and Senior Expert Advisers. This composition creates a TEAP of twenty-one to twenty-four technical and economic experts. Each TOC has at least one co-Chair from an Article 2 Party and one from an Article 5 Party. The TEAP itself has three Co-Chairs: two from Article 2 Parties and one from an Article 5 Party.

TEAP members are nominated by national governments and must be confirmed as members by a Meeting of the Parties. The criteria for the selection of TEAP members include (1) relevant technical expertise; (2) managerial and leadership skills; and (3) geographical balance in the composition of the TEAP.

TEAP members are required to disclose interests that might be prejudicial to their objectivity in carrying out their responsibilities as TEAP members. Furthermore, TEAP members must be independent experts in the sense that they must not act under instruction.

The members of the Technical Options Committees (TOCs) are selected on the basis of criteria that are broadly similar to those for TEAP members in that they must be nominated by national governments, selected for technical expertise, contribute to geographical balance within the TOC, and able to work in highly motivated expert teams. The guideline for the size of TOCs is twenty-five to forty members.

One notable feature of TEAP's early years (1989-1990) is that chlorofluorocarbons (CFC) producers were not eligible for TOC membership.

TEAP can also create "subsidiary bodies" to facilitate flexible responses to specific requests by the Parties for reports on selected issues. This approach makes it possible for TEAP to co-opt other experts with specific expertise to complement TEAP's expertise as and when required. Such subsidiary bodies are usually referred to as Task Forces and exist only for the duration of the specific task in hand. This approach has proved to be very effective, for example, in preparing the report on the 1997–99 Replenishment of the Multilateral Fund.

How it works

The work of the TEAP is directed by requests from the Parties to the Montreal Protocol for the preparation of reports on specific issues and for Full Assessments at least every four years. TEAP is currently organising its resources to prepare the 1998 Assessment with a deadline of end-October 1998.

TEAP reports present "policy relevant technical information" in the form of inputs to the policy process that drive the evolution of the Montreal Protocol. They are technical in nature with care being taken to exclude any material that might be seen as "political". TEAP usually meets formally about twice a year, but more frequent informal direct contacts among members are common.

It is also important to note that TEAP reports are drafted and approved by consensus. However, Senior Advisers to the TEAP are not eligible to vote in the process of determining a consensus on individual reports.

TEAP reports to the Parties draw on reports prepared by the respective TOCs. TOC reports are reviewed, but not revised, by the TEAP. They are published as approved by consensus of TOC members. Information presented in the TOC reports may be cascaded as inputs to TEAP reports that may present wider assessments of issues that the Parties may have mandated to the TEAP. Both TOC and TEAP documents are made available to the Parties and to the public at large. Many reports can be downloaded from the TEAP website at <http://www.teap.org>.

Why it works

The key reasons why the TEAP process works are to be found in the technical expertise, the global reach of personal and professional contacts, and the influence of TEAP and its members on relevant technical and policy issues.

TEAP and its TOCs benefit from the insights gained from continuing inter-action with a globally balanced membership and from their contacts with the wider international community of technical and economic colleagues who share a common environmental concerns.

It is essential to TEAP's effectiveness as an instrument of the Montreal Protocol that its work be based on a carefully guarded reputation for independence, objectivity, and technical integrity.

These characteristics of the TEAP allow it to serve as an effective complement to the scientific and political processes of the Montreal Protocol.

TEAP assessment process

The Parties request the Panels (Science, Environmental, Technology and Economic) to prepare full assessments at least every four years. These are key documents that require substantial efforts by the TEAP and its TOCs. These TEAP Assessment Reports present the Executive Summaries of the respective TOC Assessment Reports complemented by additional information prepared by the TEAP itself for the information of the Parties. As mentioned previously, TEAP is currently gearing up for the 1998 Assessment.

UNEP takes "ownership" of these documents which gives them substantial influence and ensures their widespread distribution. The information in the TEAP and TOC reports supports decisions taken by the Parties to the Montreal Protocol.

The Parties also request TEAP to prepare special in-depth reports on specific topics. These tend to be especially important for time sensitive issues which would not normally be covered adequately in the more formal and less frequent Assessment Reports. The Parties also request TEAP to address specific topics in Annual Progress Reports. These are substantial reports covering a wide range of issues. For example, Volume 2 of the 1997 Progress Report addressed twenty-one separate issues at the request of the Parties.

The Annual Progress Reports may also include the substantive reports of Subsidiary Bodies, such as those of the Task Force on the 1997–99 Replenishment of the Multilateral Fund and the Task Force of the Economic Options Committee on the Economic Viability of Methyl Bromide Alternatives.

Inputs into the political process

TEAP and TOC reports are technical reports. They do not interpret or offer opinions regarding possible political implications of technical assessments. The Parties request TEAP to report on their technical assessments as one of many inputs into their decision-making with respect to the Montreal Protocol.

Typically, these requests from the Parties have been in the following areas:
1) Advice on consumption and production controls of specific ODS;
2) the identification and elimination of technical barriers to ODS phase-outs;
3) assessments of the contribution of technical innovation to ODS phase-outs; and
4) assessments of the contributions of national initiatives to ODS phase-outs.

It is critical to the TEAP process that these inputs be agreed by consensus within the TEAP and/or the individual TOCs.

COUNTRIES WITH ECONOMIES IN TRANSITION

László Dobó and Lambert Kuijpers

Background

The Countries with Economies in Transition (CEIT) are the states of the former Soviet Union and those of Central and Eastern Europe. These countries have experienced massive geopolitical and economic changes over the last decade which have resulted in severe financial and administrative difficulties for both government and industry. These developments have also adversely affected the CEIT's contribution to the implementation of the Montreal Protocol. Several Country Programmes completed before 1996 for the CIS (Commonwealth of Independent States) member states reported that a phase-out by 1996 would be impossible and that there would be a delay in compliance. This was—and is—a problem for the Montreal Protocol.

At the time of drafting the Montreal Protocol, nobody envisaged the dissolution of states which were Parties to the Protocol (i.e., the Soviet Union, Yugoslavia, etc.), into a number of independent states. In the case of the successor states to the former Soviet Union, the Russian Federation, Belarus, and Ukraine continued to be Parties. The other newly formed independent states have not yet become Parties. (See Appendix for the list of the CEIT and their status as Parties.) Since the dissolution of the Soviet Union at the end of 1990, five new CIS successor states (in addition to Belarus, Russia and Ukraine) and the three Baltic Republics became Parties (five CIS member states have ratified the London Amendments, Azerbaijan, Turkmenistan, Belarus, Russia and Ukraine). Most of the CEIT have experienced difficulties in data reporting.

TEAP Study on CEIT

The Parties to the Montreal Protocol requested the Technology and Economics Assessment Panel (TEAP) to investigate the specific issues related to CEIT compliance with the Protocol. The TEAP Task Force was organized in late 1994 and made significant progress during March–November 1995. The work of the Task Force centred on collecting ODS consumption data, investigating the earliest possible phase-out dates, and identifying options that could be considered by Parties to deal with anticipated non-compliance by CEIT. It issued a final report in November 1996.

Decisions taken by the Parties in 1995 (Vienna, Austria) and 1996 (San José, Costa Rica) establish the state of non-compliance for several CEIT countries and urge them to take adequate steps to achieve compliance at shortest notice.

The Fourth Meeting of the Parties to the Montreal Protocol in 1992 in Copenhagen, strengthened the Protocol by adjusting the phase-out dates of halons to 1994 and of chlorofluorocarbons (CFCs), carbon tetrachloride and methyl chloroform to 1996. It also added to the list of controlled substances (by an Amendment) hydrochlorofluorocarbons (HCFCs) and methyl bromide. This strengthening of the Protocol was justified both by the observed serious depletion of the ozone layer which was occurring faster than previously anticipated and by the rapid progress in research, development, and application of substitute compounds and/or processes. The progress of substitution and phase-out of ODS in the developed countries during 1993 and 1994 demonstrated that the phase-out deadlines of the Copenhagen adjustments and Amendment could be realised by the developed countries.

Compliance Challenges for CEIT

During the years 1993 and 1994, it became increasingly obvious that a number of countries in Central and Eastern Europe would have difficulties in meeting the goals and obligations of the Protocol. Non-compliance was considered highly probable in those countries which became independent states after the dissolution of the former Soviet Union. The successor states to the former Federal Republic of Yugoslavia were classified as operating under Paragraph 1 of Article 5, therefore compliance issues are not as pressing as for other Parties.

Many of the successor states of the former Soviet Union have currently acceded to the Protocol. However, a number of them remain non-Parties. Information on activities or data of production, consumption, and substitution/phase-out of ODS were—except in some Central/Eastern European countries like Bulgaria, the Czech Republic, Hungary, Poland, Romania and Slovakia—extremely scarce or non-existent. With the dissolution of the Soviet Union, the official reporting of production, export and consumption data of the largest producer, exporter and consumer in the region was terminated in 1990.

The scale of this anticipated non-compliance with the developed country 1996 phase-out of the controlled substances (Annex A/B) has presented the Parties with a

challenge to achieving the objectives of the Protocol. The policy problem is particularly complex for the following reasons:
1) For most of the CEIT that are Parties, the Protocol was ratified by the former Socialist States and not by the current States.
2) Not all CEIT, are Parties, particularly several CIS states.
3) Some of the CEIT may qualify as Article 5(1) Parties and therefore benefit from grace periods regarding ODS phase-outs (Georgia, Romania and the successor states to the former Yugoslavia are or were (re-)classified as Article 5(1) Parties).
4) Not all the CEIT (several CIS member states in particular) were fully aware of the implications of the 1996 phase-out, or are aware of the consequences following the January 1996 phase-out.
5) CEIT, and particularly the CIS member states and the Baltic countries, face binding constraints on institutional capacities, information on alternative technologies, and financial resources.

The financial constraints the CEIT (particularly CIS member states and the Baltic countries) face in addressing the phase-out of ozone-depleting substances are currently one of the most important factors behind non-compliance with the Protocol.

When the Ozone Secretariat enquired at the United Nations Legal Office on the status of the countries of the former Soviet Union, the response was the following:

> A distinction should be made between Belarus and Ukraine on the one hand, and the other Republics on the other. Belarus and Ukraine were already parties, in their own right, to the Vienna Convention and the Montreal Protocol. Accordingly, their position with regard to these [treaties] remains unaffected.

> As far as the other Republics of the former Soviet Union are concerned, they are treated as newly independent States and legally distinct from the former Soviet Union. Under the practice of the Secretary-General as depository of multilateral treaties, they could become Parties to the Convention and the Protocol only pursuant to an explicit expression of consent to do so by their respective governments. Such consent can be expressed either by (a) depositing an instrument of accession expressing a consent to be bound by the Vienna Convention and the Montreal Protocol, or (b) by explicitly succeeding to these instruments which were previously applied on their territory. The Secretary-General's practice does not include the concept of an "automatic succession". It is for a newly independent State to decide whether or not it should become a Party to any treaty deposited with the Secretary-General. (see report UNEP/OzL.Pro/ImpCom/14/2, 4 June 1996)

The Technology and Economics Assessment Panel's Task Force on CEIT has worked constructively with the affected countries, and their efforts have contributed significantly toward achieving several important goals:
1) Compiling and presenting a reliable estimate of ODS production and consumption data in almost all CEIT, especially in all successor states to the former Soviet Union, from 1986 through 1995/1996. These estimates were the first and only data after the Soviet Union's last 1990 report. Subsequently, the

official reports of the Russian Federation, Belarus and Ukraine confirmed the validity of the Task Force's estimates for 1990-1994.
2) Producing the first estimate of the incremental costs of phasing out ODS in the CEIT.
3) Accelerating the ratification process in several countries (by efforts undertaken in 1995 and 1996, including direct contact with several countries that had/have not ratified).
4) Assessment of specific problems and difficulties in non-Party countries related to ratification, in particular development of a better understanding of the infrastructure problems and the difficulties in international communications.
5) Assistance to the Implementation Committee regarding ways to deal with non compliance by certain countries.
6) Assistance to non-Party (CIS) countries by providing detailed information and suggestions (in English and Russian). Non-Party CIS states specifically requested unambiguous guidance on how to become a Party, what their specified obligations under the Protocol are, and how to obtain financial and expert support to fulfil these obligations.

It is important to note that the Technology and Economics Assessment Panel Task Force emphasised these issues in its November 1995 report and asked for further guidance from the Ozone Secretariat and from the President of the Implementation Committee. While the responses were correct from a legal standpoint, they did not provide guidance which would assist the development of a simple and pragmatic approach to assisting CEIT. Therefore, the Task Force compiled basic information regarding the situation in CEIT, and developed its own suggested practical steps to ratification and methods on how to obtain financial and expert support, which were included in its' 1996 report.

Despite these accomplishments, there remain a number of issues the Task Force has not yet been able to resolve:
1) Ratification by Parties and non-Parties of the Protocol and the London Amendment if they perceive the financial obligations under the Multilateral Fund as prohibitive.
2) How to address trade in ODS between Parties and non-Parties (particularly between the Russian Federation and other CIS successor states);
3) Co-operation on data reporting with new Parties, and not only with non-Parties, particularly on consumption data (related in part to general difficulties in data reporting, and to the non-availability of data over the period 1986–1996).
4) Adequate assistance in identifying conversion projects in Party and non-Party countries for submission to the Global Environment Facility (GEF), following ratification of the London Amendments;
5) Adequate monitoring of project execution in those countries that have projects approved under the GEF.

Equity issues

Two situations in particular bear on the fairness of inflexibly applying the terms of the Protocol. First is the financial obligations of CEIT, and particularly the levels of contribution by CIS and Baltic countries to the Multilateral Fund. Given the current situation in these countries, their status may be further evaluated by the Parties. Second, bilateral or multilateral assistance from other Parties could provide the support needed to aid these countries achieve compliance.

The following factors bear directly on these issues.

1) *Non-participants in the 1987–1990 Protocol negotiations*
   Since the new CIS and Baltic countries did not exist at this time, they were not able to participate in the most significant negotiations of the Montreal Protocol in 1987 and 1990 (the basic agreement and London Amendments). They perceive that the obligation that would be placed upon them on ratifying the London Amendments as not "legally binding", but rather as an injustice, since the international community is trying to force them to proceed as if they were/would have been independent states and equal negotiating Parties during 1987–90.

2) *Article 5(1) status application*
   The new CIS states and Baltic countries have no common viewpoint regarding the application for (re-)classification as operating under Article 5 Paragraph 1. Rather, they question whether the list of developing countries in the Montreal Protocol is legally binding, since the new CIS states and Baltic countries did not exist at the time that the list was made. Furthermore, they question the rule that attributes the same status to successor states as the country that has been dissolved.

3) *The ratification of the London Amendment and subsequent payments to the Multilateral Fund*
   Because ratification of the London Amendments implies payments to the Multilateral Fund, ratification becomes prohibitive by the new CIS and Baltic countries since their GDP and financial situation does not allow them to make the required contributions. Although these states realise that payments to the Multilateral Fund can be postponed (for the CIS countries, as addressed in Decisions VII/17, 18 and 19 by the Montreal Protocol Parties in Vienna, 1995) and that GEF support would be applicable, their national laws forbid most of them to follow such practices (which the CIS and Baltic countries regard as valid for all international treaties).

4) *The large number of International Conventions and Protocols*
   The international community is urging the new states to ratify a large number of International Conventions and Protocols at short notice. The human resources to deal adequately with the ratification are often lacking while their budget situation is causing delays.

Several countries have ratified the Montreal Protocol but not the London Amendments (e.g., Estonia, Georgia, Latvia, Lithuania, Moldova). Yet they are listed as contributors to the Multilateral Fund with payments in arrears during 1994–1996. These countries consider this listing to be more than a mistake: they suspect it reflects international pressure to ratify the London Amendments without giving the adequate time to discuss this step. Again, they consider such pressures to be against "international law".

Ratification Options

The new CIS member states and Baltic countries identified the following options for achieving what they would consider a "valid" status under the Montreal Protocol (these options were extensively discussed during a workshop in Riga in November 1996, and reported in the TEAP Task Force Final Report).

1) *Ratify the Montreal Protocol only, and not its Amendments*
   This would open opportunities to attempt compliance by replicating the same actions already completed in other countries, and to finance conversion projects via bilateral agreements and loans.
2) *Apply for Article 5(1) status-reclassification*
   Successful application would waive contributions to the Multilateral Fund, but uncertainty about achieving reclassification is disadvantageous since it would postpone any possible conversion projects for several years.
3) *Submit project proposals to the GEF without ratifying the London Amendment*
   Once project approvals are granted, ratification of the London Amendment can be considered on the basis of whether the package of approved projects provides sufficient confidence that compliance could be achieved.
4) *Ratify the London Amendment, postpone payments to the Multilateral Fund, and submit project proposals to the GEF*
   This option—which was selected in a number of earlier cases, e.g. for Russia, Belarus, Ukraine—is subject to objection, based on interpretation of national law.
5) *Ask for special status after ratification of the London Amendment*
   Several countries would prefer to obtain a special—Parties' endorsed—status concerning their contributions to the Multilateral Fund, so that they can proceed, without facing conflicts with their national regulations.

The way ahead

While progress, including a 1996 phase-out, has been made in several CEIT, particularly in Central Europe, significant steps still need to be undertaken in many CIS member states and in the Baltic countries. It should be emphasised that substantial differences exist among the CEIT/CIS states with respect to their CFC consumption, their progress toward a CFC phase-out, the status of their Multilateral Fund contributions, and their significant differences in Gross Domestic Product (GDP).

For many of the CEIT outside the CIS and Baltic region, the procedures, avenues for financial assistance, and obligations under the Montreal Protocol are now well outlined. For most of the CIS member states and Baltic countries, the situation is moving towards a resolution, although a "definitive" solution has not yet been achieved. Follow-up activities by the countries concerned, in addition to participation in OEWG Meetings and the Meetings of the Parties are required.

The most important issue precluding ratification of the Protocol and its Amendments by CEIT and CIS states (particularly the latter) is the fulfilment of their financial obligations to the Multilateral Fund, since most of them are classified as developed countries under the Montreal Protocol. Virtually all of those which have ratified the Montreal Protocol, face overwhelming difficulties that at present preclude their contributing to the Multilateral Fund.

The outcome of the Preparatory Meeting in San José, Costa Rica just before the Eighth Meeting of the Parties related to financial contributions, makes it unlikely that a special financial status for certain CIS and the Baltic countries will be possible in the near future. Although Georgia was successful in applying for the Article 5(1) status, any further application will encounter substantial difficulties. It is highly improbable that such applications would be considered during the period 1997–1999. Further progress in the near future regarding financial obligations (either via bilateral and multilateral contacts, or via small regional meetings) is difficult to judge at present.

The TEAP Task Force observed that the main barriers to expediting phase-out progress in the CIS states and the Baltic countries are as follows:
1) The perception of many CIS and Baltic countries that they cannot ratify the London Amendment since it would imply financial obligations which they perceive as impossible to fulfil.
2) Difficulties in ways and means of advising countries regarding their financial obligations (several CEIT/CIS countries have ratified the London Amendment, others have not).
3) The difficulty of identifying the official entity responsible for the ratification process in many CIS countries.

Because overcoming these barriers requires significant effort and innovative solutions, the Task Force recommended that the responsible entities undertake direct efforts in each country, such as
1) preliminary country programme preparation;
2) assistance for initial data reporting in those countries that have so far not ratified;
3) preliminary project proposal preparation of conversion projects to stimulate progress towards complying with a phase-out;
4) meetings with the appropriate officials from responsible ministries either in small regional meetings or via a country by country approach;
5) inviting non-Parties to OEWG Meetings and to the Meetings of the Parties.

Three of the issues mentioned above have already been addressed by:

1) inviting non-Parties to OEWG Meetings and to the Meetings of the Parties, with financial support from bilateral contributions;
2) assisting in the preparation of preliminary Country Programmes for certain CIS countries;
3) recommending holding of an intergovernmental meeting with high-ranking officials from the responsible ministries of several countries.

It is anticipated that in most CIS states and in the Baltic countries non-compliance will occur over the next three–five years. This information was obtained from statements by the Russian Prime Minister, from Country Programmes, and from discussions with experts in these countries. It was also reported to the Meetings of the Parties in 1995, and subsequently. The fact that several of the twenty-seven countries involved have not yet ratified the Montreal Protocol makes widespread non-compliance inevitable.

The report also presents a number of options that Parties may wish to consider in order to promote an ODS phase-out in the CEIT. It may be important to emphasise the interdependence of all CIS and Baltic countries in considering non-compliance. Finally, a number of options were developed to assist with compliance. These include:

1) Encourage non-Party CIS states to ratify the Montreal Protocol and the London Amendment, by inviting them to OEWG Meetings of the Parties and to provide those countries with a concise information package on how to proceed. This package should contain guidelines regarding the ratification procedure and data reporting in Russian. It should also contain a short description of the necessary elements of a Country Programme, also in Russian.
2) Request the TEAP to cooperate with the Russian Federation in order to establish data on imports of ODS by CEIT non-Parties from the Russian Federation (which could be cross-checked with the data the country submits as part of the ratification procedure) and to forward this data to the non-Party countries. Parties may also consider to request UNEP to defer financial obligations to the Multilateral Fund for those countries which ratify and submit data on consumption, until such time as their status has been clarified; and to inform these non-Party countries accordingly.
3) Fund all ODS projects in CIS states (non-Article 5(1) and possible Article 5(1)) via the Global Environment Facility (and other financing means such as multilateral equity funds) in order to accelerate their compliance with the Montreal Protocol.

APPENDIX
List of Countries with Economies in Transition and Status as Parties

Albania has initiated a process leading to ratification of the Montreal Protocol. After ratification, Albania will automatically be classified as operating under Paragraph 1 of Article 5.

Armenia has not ratified the Montreal Protocol and would be inclined to ask for Article 5(1) re-classification from the Parties to the Montreal Protocol.

Azerbaijan ratified the Montreal Protocol and its London and Copenhagen Amendments in June 1996.

Belarus ratified the London Amendment by June 1996.

Croatia's Country Programme was completed in May 1996.

Estonia ratified the Montreal Protocol in October 1996.

Georgia ratified the Montreal Protocol in March 1996 and successfully applied for (re)-classification as an Article 5(1) country to the Eighth Meeting of the Parties in San José, Costa Rica, in November 1996 (as adopted via Decision VIII/25).

Kazakhstan has made enquiries regarding its financial obligations to the Multilateral Fund. No further information regarding Kazakhstan's intentions has been received.

Latvia's situation of non-compliance was addressed by the Parties in November 1996. Latvia has not yet ratified the London Amendments.

Lithuania's non-compliance was addressed by the Parties in November 1996. Lithuania has not yet ratified the London Amendments.

Macedonia (former Republic of Yugoslavia) completed its Country Programme in July 1996.

Moldova ratified the Montreal Protocol at the end of October 1996. It applied for (re-)classification as an Article 5(1) Party; however, approval is uncertain.

Mongolia ratified the Montreal Protocol and its London and Copenhagen Amendments in March 1996.

Poland ratified the London and Copenhagen Amendment on 2 October 1996. A GEF grant to phase out ODS has been approved.

The Russian Federation has informed the Ozone Secretariat and the Implementation Committee of its non-compliance with the Montreal Protocol. The 8th Meeting of the Parties in 1996 noted that considerable progress has been made by the Russian Federation. However, disbursement of financial assistance should continue to be contingent on progress toward achieving compliance.

Tadjikistan ratified only the Vienna Convention in June 1996.

Turkmenistan ratified the London Amendments in early 1994. Its Country Programme was initiated with GEF support.

Ukraine ratified the London Amendments and contributed U.S.$ 785,600 to the Multilateral Fund. It ratified the London Amendments on 15 November 1996.

Uzbekistan has begun preparation of its Country Programme with the support of the GEF. It has not yet ratified the London Amendments, however.

IMPORTANCE OF THE TEAP IN TECHNOLOGY COOPERATION

Sally Rand and Lalitha Singh

Previous papers focused on the importance of the Technology and Economic Assessment Panel (TEAP) and the technology assessment process in the context of the political process. TEAP supports not only the political process of the Montreal Protocol but also plays a large role in the transfer of technology. The objective of this paper is to describe the contribution of TEAP to international technology co-operation and the implementation of technologies that have been identified throughout the process. This additional channel of support for the replacement of ozone-depleting substances has been fundamental to the success of the Montreal Protocol and should be carefully considered for other global environmental issues.

The TEAP has been successful in this task through:
1) selection of best technical experts;
2) members "professional growth" through the work;
3) ensuring objectivity via technical and geographical diversity; and
4) receipt of support from a wide range of organizations.

Individual TEAP members have been instrumental in implementing many TEAP recommendations because participants are:
1) members of industry technical and policy committees;
2) advisors to national governments;
3) participants in government and corporate policy teams;
4) corporate decision makers; and
5) technology inventors and champions.

The selection of highly motivated, independent experts as TEAP members contributes to the success of the Montreal Protocol because they:
1) formulate self-effecting solutions;
2) are respected technical ambassadors;
3) develop extensive international and domestic networks; and

4) are themselves leaders in their fields.

Members of the TEAP and TOCs provide benefits to their industries within their own countries and in other countries by becoming technology ambassadors. The process facilitates interaction of experts from other countries and across industries. Sharing information accelerates transfer of global technical and policy developments to each members company and technical community. Sharing experience on alternatives builds confidence in choices of technical options. As a result, members are looked to as advisors to governments and to industry. Similarly, they bring to the TEAP and TOCs information about technical innovations in their own countries.

Through their international perspective, they become inventors, innovators and technology champions. As a result, members have access to leaders and decision-makers both within their own countries and internationally. It was this "out of the box" thinking that resulted in no-clean technology for electronics manufacturing. The electronics industry used large quantities of CFC-113 to clean circuit boards following soldering operations. Their initial objective was to find an alternative cleaning technology. However, by looking more comprehensively at their manufacturing processes, they discovered that by changing the flux used within the solder joints, and by modifying the soldering operation itself, they were able to produce solder joints that did not have to be cleaned at all after soldering. This resulted in improved quality of solder joints, reduced the reject rate for boards following the soldering operation, and greatly reduced operating costs by avoiding the need to buy expensive solvents, dispose of waste, and maintain the equipment used for the cleaning operation.

This is also an excellent example of self-effecting change from within an industry brought about by technology innovation. TEAP and TOC members are independent volunteers, who individually and collectively bring great value to the Montreal Protocol process. Because of this, these organizations could not be replicated, nor provide the same value to the Protocol process if these were paid positions.

The TEAP process has created literally hundreds of technology ambassadors who have been able to move out into the world via the technology assessment process in order to promote new technologies and facilitate their implementation. This is possible because TEAP started with the best technical experts thanks to its recruitment policy of respected, independent volunteers. TEAP has created an environment where people can grow with the work. The combination of disciplinary, geographical, cultural diversity, and common purpose has created a unique environment in which to train people in the art of international technology cooperation. TEAP members are not necessarily experts in the process of the Montreal Protocol or in global environmental issues but they are technical experts in industrial applications of ozone-depleting substances who are dedicated to helping their industries and nations meet the requirements of the Montreal Protocol. TEAP provides a forum for these experts from around the world to join intellectual forces for a shared environmental goal.

Organisational support has also been very important in this process. Most members serve as volunteers. Their time and travel support translate into financial contributions to the Montreal Protocol. Each TOC and the TEAP produce substantive annual reports and conduct special studies as requested by the Parties to the Protocol. Members spend significant amounts of time—both their employers' time and their own—preparing these reports, coordinating with TEAP colleagues, and attending meetings. These are not insignificant contributions. Currently, the only costs to the Montreal Protocol for TEAP activities is travel support for members from developing countries, and that funding is barely adequate to accomplish the demanding workload. Without this in-kind support from members' employers and governments, the TEAP could not exist in its current form. If these were paid positions, the process would not attract the personal commitment found among the volunteer TEAP members.

TEAP and TOC members are influential not only in advising the Parties but also within their own professional communities. TEAP members participate in technical and policy committees that set standards, trends, and the pace of change within industrial sectors. They are influential advisors to national governments. A good example comes from our work for national governments. When faced with various sector issues or questions as to the needs and positions of various industries, TEAP and TOCs reports and the views of individual TOC members are looked to as objective and comprehensive. Thus participants become influential across their respective communities.

TEAP and TOC members are also corporate decision-makers. Members are owners and senior managers of their companies, and they bring information from the Montreal Protocol back to their companies and influence their decisions.

The atmosphere in the TOCs is unique because substitutes are identified, refined and even invented. There is the case of Richard Nesbaum, who was the original inventor of the widely used CFC-based medical sterilant. He became a member of the Aerosol and Medical Sterilant TOC and actually went on to invent a substitute that became an effective drop-in replacement. TEAP has also produced technology champions. A good example is the Solvents Committee which devised the idea of no-clean technology for cleaning printed circuit boards. This is complete out-of-the-box thinking which will not substitute, but eliminate the need for solvents. These are just a few examples of how individual members have become very influential in their professional communities.

In summary the opportunity for self-effecting change is invaluable. The understanding and confidence been gained through the TOC assessment process have created technical ambassadors who have proceeded not only to champion technologies, but to push the transition on a pace that would not otherwise have been achievable. The TEAP and the TOCs have created significant networks for people including unique access to policy-makers and government leaders. Finally, individual members have shown tremendous leadership in championing both the Montreal Protocol and the potential to complete this process.

It is clear that the rapid and successful implementation of the Montreal Protocol would not have been possible without the TEAP process and the TOCs. It is certainly a process that money could not buy. The cost of assembling a paid group of experts of the size and stature of the TEAP and TOCs would be prohibitive. But perhaps most importantly, it would not be possible to replicate the passion, commitment and momentum of TEAP and TOCs by employing paid staff—thus is the nature of volunteers who believe in their cause. The TEAP has influenced the political process through its success in establishing a credible, objective and transparent technical assessment process, and greatly contributed to the implementation of the Protocol through technology cooperation.

SCIENTIFIC OBJECTIVITY, INDUSTRIAL INTEGRITY AND THE TEAP PROCESS

Lambert Kuijpers, Helen Tope, Jonathan Banks, Walter Brunner and Ashley Woodcock

Introduction: *Dr. Lambert Kuijpers*

The principles of scientific objectivity and industrial integrity are critical to the TEAP's ability to provide useful policy-relevant, technical information to the Parties to the Montreal Protocol. The Parties rely on the TEAP's ability for objective and balanced information upon which to base their decisions. Reports are developed through a consensus approach and this leads to the quality technical data on which the parties can rely. TEAP and TOC members are charged with the responsibility of acting independently—without instruction from employers, including governments— and participants act in a voluntary capacity. In many cases members are drawn from industry with direct experience in the use of ODS and their alternatives. It is important to have individuals with the integrity to remain independent despite the funding they receive from their sponsoring organisations or companies.

There have been many successes and challenges in the TEAP's efforts to maintain high standards of scientific objectivity and industrial integrity. One of the successes has been the trust developed in technical aspects of the Montreal Protocol process. The Parties trust the information they receive, which allows them to move forward confidently. Another success has been the building of a cooperative spirit between all stakeholders to find solutions to the problems created by the consumption of ozone-depleting substances. TEAP has been outstandingly successful in avoiding provision of political comment, giving only technically based advice to the Parties. Most importantly, false and misleading claims cannot be maintained in the TEAP environment which is based on scientific objectivity and industrial integrity.

Despite the success of the TEAP, there are new challenges to its objectivity and integrity. TEAP and TOC members are being challenged to resist new lobbying efforts and instructions by governments. Another challenge is that, with large business interests having a stake in the outcome of the TEAP reports, large numbers of people desire membership. The problem is that such large committees become unmanageable, lead to duplication in areas of expertise, and harbour the potential for factions to develop with their own agendas. The challenge therefore, is to keep the TEAP and the TOCs to a manageable size, while maintaining the necessary mix of expertise and geographic representation. Members have typically spent their careers in technical and academic arenas and can be unprepared to cope with personal attacks or innuendo designed to undermine the credibility of the process as well as their own. The challenge remains to ensure the highest levels of scientific and technical objectivity and industry integrity in what becomes, at times, a hostile environment, and to produce the best consensus-derived technical information for the Parties.

The following brief descriptions by some of the TOC co-Chairs will exemplify how objectivity and integrity, and TOC performance, are being maintained in the face of pressures coming from those who pursue singular agendas.

Methyl Bromide Technical Options Committee: *Dr. Jonathan Banks*

The Methyl Bromide TOC (MBTOC) was the last TOC to be established. It deals with methyl bromide alone, which was the last ODS to be recognised and controlled by the Montreal Protocol. The committee was set up around 1994. It was originally a very large committee with up to sixty-five members. It had to be that large in order to accommodate the very diverse interests and opinions that prevailed at the time on methyl bromide.

Its composition included manufacturers of methyl bromide, a number of scientific methyl bromide experts from various fields, and environmental activists with expertise in alternatives to the use of methyl bromide. The diverse spectrum of representation created a fragile system in which objectivity was to be very important in order to reach a successful, consensus-based outcome.

Methyl bromide control brought some very specific problems to the Protocol. First, methyl bromide is an agricultural chemical—the first time the agricultural sector came under the scrutiny of the Montreal Protocol. The industrial sector—agribusiness—was not accustomed to the process. Furthermore, agricultural users of methyl bromide are a risk-averse and conservative user group. An additional problem is that, unlike other industrial sectors affected by the Montreal Protocol, the methyl bromide industry produces no alternatives and therefore has no business interest in alternatives. As a result, the producing industry has no stake in advancing an alternative chemical or technology and is very much on the defensive.

It was experimental to include such a large number of chemical manufacturers and suppliers in a TOC. During the early formation of the technical options committees, participation by chemical producers was prohibited since it was believed that with no business interest in alternatives, objectivity would not be possible. Eventually, it became evident that alternatives would be a combination of new chemicals and not-in-kind, and participation by chemical producers was permitted. When the MBTOC was formed, the chemical producers and suppliers were permitted as members from the start. This produced a very diverse committee. Nevertheless, the committee was able to work toward a consensus document which guided the Meeting of the Parties in Vienna in 1995.

"Would the report have been any different if MBTOC had not included a considerable number of representatives from the industry that very produces and depends on methyl bromide?" According to Jonathan Banks, however, the inclusion of the producing and suppliers industries was important, and resulted in an authoritative and soundly based report.

Since the production of the 1994 MBTOC Assessment Report, problems have arisen. As the viability of alternatives has been demonstrated in various places around the world for a wide range of methyl bromide applications and for a number of important crops, the producing industry has become increasingly concerned about the long term viability of their business. As a result, the committee is now polarised into two groups—one defending methyl bromide and minimizing the importance and significance of the not-in-kind alternatives, and another strongly and convincingly promoting the successful methyl bromide alternatives. With the emergence of a powerful and well financed pro-methyl bromide group, the MBTOC has become less constructive and the emphasis on well-proven alternatives is being diluted by a well financed publicity campaign aimed at discrediting the alternatives and those who promote them. During meetings and debates on methyl bromide, you are likely to hear people proclaim that there are no alternatives to methyl bromide— but this is clearly not the case. It has become increasingly difficult to reach consensus within MBTOC and much of the debate is now taking place outside the committee meetings, and is being taken to the farming community and government legislators through a well financed private sector lobbying and publicity campaign. This carefully orchestrated criticism of MBTOC activities taking place outside the Protocol makes it increasingly challenging for the committee to provide the Parties with objective policy-relevant information. This situation is destabilizing the MBTOC and represents a threat to the total Montreal Protocol process.

The MBTOC is now being restructured into a smaller group, with a new emphasis on balanced representation and a focus on the technical and economic viability of methyl bromide alternatives. During this restructuring, great care is being taken to maintain a balance of opinion in order to maintain objective debate. However, it is important that the debate represent an honest exchange of information and not a competition of financial wherewithal to wage lobbying and publicity efforts. The MBTOC will continue to produce credible reports with integrity. It will continue to focus specifically on alternatives to the use of methyl bromide as opposed to focusing mainly on those situations in which methyl bromide must continue to be used.

In summary, the Methyl Bromide TOC has worked well in the past, but has become a target of aggressive business interests intent on dominating the debate.

The Parties have recognized the recent problems and their causes and have supported the restructuring in order to return objectivity and integrity to the process.

Aerosols Technical Options Committee: *Dr. Ashley Woodcock and Dr. Helen Tope*

Dr. Woodcock is a physician looking after patients with asthma and chronic obstructive pulmonary disease (COPD). Metered-dose inhalers (MDIs) are the last significant use of CFC in developed countries. These products must be phased out, and this presents a big problem. It is a big problem because there are a large number of patients involved around the world, and physicians must maintain the primary principle of patient safety. By way of perspective, there are approximately 500 million MDIs for seventy million patients. It is also a complex problem because a wide range of companies manufacture MDIs, ranging from large multi-nationals which primarily manufacture name brands, to small national companies which primarily manufacture generic MDIs.

The development of alternatives has been complex and technically challenging. However, alternatives have been developed; phasing out this use of CFCs is technically feasible. Companies have probably spent one billion dollars so far to produce CFC-free metered-dose inhalers. This has involved extensive safety and efficacy testing. The new inhalers must meet much more stringent regulatory requirements than the original CFC MDIs ever had to achieve. There is a complex and uncertain patent position. It is likely that the number of companies that will manufacture MDIs in the CFC-free formulation will be fewer than those currently available since intellectual property has expired for all the CFC MDIs. It is a very complex situation; a large number of factors must be taken into account. What we are trying to do as a TOC and the TEAP is to achieve a correct balance between an early phase-out to protect the environment and protecting patient's safety.

The TOC has been a cohesive group of environmental experts, physicians, and technical experts. The technical experts, although employed by the pharmaceutical companies that manufacture the CFC MDIs, have been remarkably independent in their positions and in the information that they have contributed. They do not represent their companies. The process would have been impossible without them and they have acted with integrity throughout this process.

The TOC has had vigorous, and at times, heated debate, but has always managed to come to a consensus. Whereas for Mrs. Thatcher consensus was a "dirty word" as it meant abandoning your principles, John Major, on the other hand, believed that it was a matter of taking all opinions; not summing up down the middle, but doing the right thing. That is what the TOC and the TEAP tried to do.

The process has gone very smoothly from the technical perspective because, until 1996 or so, there has been no commercial advantage for any of the companies. The first CFC-free MDIs came out in 1995 and only by one company initially. No commercial advantage, no tension, no profit. Once alternatives were launched, tensions started developing. When policy and profit diverge, tensions will develop.

The innovators and patent holders want a fast transition. Companies who are changing over more slowly need a slower transition. There are also companies that are not producing any CFC-free MDIs; they want to go on producing CFC MDIs forever and reject any transition. All these companies now employ lawyers and lobbyists, with the lobbying being carried out both at the Party and TEAP levels. However, unlike the MBTOC, the tactics have not been as intensed or personal. The companies have lined up physician and patient groups, sometimes overtly, sometimes covertly. Pressure is developing in this transition process, but the TOC believes that these pressures are almost entirely commercial, not related to technical issues. The TOC also believes that it is very difficult for those of us in the process who are not familiar with these pressures to resist them. But we will. The challenge is to continue to provide objective, independent information to the Parties so that they can achieve a rapid transition in the spirit of the Montreal Protocol. As a practising physician, I would like to tell you that we want to achieve that without compromising patient safety, wherever they may live.

Halons Technical Options Committee: *Dr. Walter Brunner*

The examples above describe a rather difficult process or one that has the potential to become difficult in time. Altough, the halon phase-out process has not been very difficult in the past, it may turn a little difficult in the future. Halons were the last ones in and the first ones out—the last of the substances to be included in the original protocol with the most lenient initial control schedule, and the first to be eliminated. They were not originally considered on the list of Montreal Protocol substances ten years ago. This is because they were for fire protection, and since lives are at stake the Parties were reluctant to include them in the initial Protocol. They were added during the final negotiations. The original Protocol controls were for only a cap at the 1986 production level. Three years later, the Parties further restricted controls and enacted a production freeze in the year 2000. However, before the time the first control even became effective, the freeze in production was advanced to 1994.

It was in some ways difficult to be first, and in other ways easy. It was difficult to get the industrial community used to a new process and especially to such a rapid process. It was easy because of the willingness of people to work constructively. Professional societies in the sectors took the lead.

The fire protection profession is accustomed to operating under the fundamental concept of assessing risks as a means to provide effective solutions. It does not merely sell products, it provides solutions. Reducing risks means that you have to take into account fire damage, property damage and damage to people. The new challenge was to include "environmental aspects" into its assessment of risk. The fire protection community, with respect to the Montreal Protocol, may have benefited from large European accidents such as the runoff from a fire polluting a river in Switzerland and a similar problem in France, not to mention the terrible chemical incident in Bhopal, India. The fire protection industry has begun to realise that environmental considerations are a part of fire protection as well. The solution to halon use also represented an opportunity for the fire protection community to strengthen the role of fire protection engineering in fire protection—a more thoughtful approach than merely selling relatively cheap, effective and widely-available solutions.

An additional advantage was that the fire protection community is rather small and closely knit—even when considered on a global basis. As a result, there were a limited number of players, with the chemical agent manufacturers only a small part of the community. Equipment manufacturers and distributors are the principal interface between halon producers and the main users (largely the military and the oil and gas industry). These large users were very important and influential players in the Halons Technical Options Committee. They are owed a big tribute. Without their contribution and commitment to ozone layer protection, such speedy success would not have been possible.

Since the fire protection community was providing protection and not product, it was easier for them to find and adopt new technical solutions. It was not even necessary to have the new agents available immediately. The business of providing solutions offered additional business opportunity because of the long-term relationship with customers.

An important ingredient in the rapid success was that the users would not lose their capital investment. Although the production of halon was being phased-out, there were no restrictions on its re-use—and important uses for which there were no alternatives could be maintained. Halon banking, as the term was coined by the committee, was necessary in order to continue critical uses for certain important applications. Equipment manufacturers recognised a new market opportunity. They could provide high-quality service for existing systems, which became much more important in a world of limited halon availability. The halon sector was also fortunate in that halon represented only a small proportion of the total CFC market, and only a very small portion of the total sales for most chemical producers, with one exception. Overall, the halon market was not that important to the large chemical producers.

However, success has created a new potential problem. As chemical alternatives are becoming available, producers are seeking to accelerate their implementation and looking for newer, bigger market shares. As a result, they are pressuring the fire protection industry prematurely to retire and decommission important halon systems. As the sizeable investments made by large former halon users in new not-in-kind alternatives (such as water mist and water spray) begin to show promise, and poised to capture large portions of the chemical replacement market, chemical producers are eager to capture this market before the environmentally preferable alternatives become widely commercialized. It is therefore important that regulators not confuse business interests with environmental concerns.

GLOBAL BENEFITS AND COSTS OF THE MONTREAL PROTOCOL

James Armstrong

Introduction

Environment Canada commissioned this study on the global benefits and costs of the Montreal Protocol as part of Canada's contributions to the Tenth Anniversary Meeting of the Parties to the Montreal Protocol. The world had a difficult choice in 1987, between the costs of protecting and restoring the ozone layer and the costs of doing nothing and living with the consequences. The occasion of the Tenth anniversary of the Montreal Protocol is an appropriate time to look back and consider whether the premises on which the Protocol are based were correct, and whether the benefits of the Protocol outweigh the costs.

The study was done for Environment Canada by ARC-Applied Research Consultants of Ottawa, a company specializing in socio-economic research and with extensive experience with the socio-economic impacts of the Montreal Protocol. The study was prepared in close consultation with an international panel of experts, which included members of the Technology and Economic Assessment Panel (TEAP) and the Chairs of the other assessment panels.

The main conclusion of the study is that the benefits of the Montreal Protocol significantly outweigh the costs associated with its implementation. Economic benefits exceed costs by some $224 billion (net present value, discounted over the time frame of study). This does not include the health benefits that global society will realize, including significantly fewer cases of skin cancer, cataracts and deaths resulting from skin cancer.

This paper provides an overview of the study, including the methodology used in the study, the benefits and costs that were identified, the challenges associated with doing a study of this magnitude, and the conclusions that were reached.

Methodology

The time frame that was eventually selected for the study was the period 1987–2060. This is the most appropriate time frame, for a number of reasons. Firstly, it matches up with the impacts of the intervention being assessed—2060 is approximately the date at which the ozone layer is expected to return to pre-1980s levels. Secondly, this time span corresponds roughly to the expected life span of many people who were born when the Protocol was signed. Thirdly, it reflects the time during which the effects of ozone depletion in the absence of the Protocol would have been most severe.

Sensitivity analyses of the time frame were done as part of the study. In addition to looking at the effects of time frames ending in 2050 and 2070, the consultants also considered the impacts that would have been associated with a twenty year delay in implementing the Protocol (i.e. not starting until 2007, rather than 1987).

The study is based on existing information, obtained from a variety of sources. Every effort was made to ensure that the information used was generally accepted. The International Advisory Panel was of great assistance in this regard. The Panel also helped to facilitate access to regional and national information.

The evaluation of the health impacts of ozone depletion utilized a comprehensive approach to quantifying UV-B exposure of different populations. The approach took into account differences in latitude, population characteristics (e.g. skin colour, life expectancy, economic development) and exposure (e.g. religious practices relevant to clothing).

In measuring the costs associated with the Protocol, the study focused on "real resource costs". These are the economic costs to society of additional resources needed to convert from ozone-depleting substances (ODS) to other substances and processes. These costs can include capital costs, research and development costs, and continuing energy, labour and material costs of alternatives and processes that are more costly than the ones they replaced. Real resource costs are borne by all sectors of society. Some are incurred immediately, while others are incurred over the whole time period to 2060. Continuing costs such as energy, labour and material costs are calculated on a yearly basis, and added up over the time period.[1]

All the annual benefit and cost estimates are discounted to 1997 net present value, using a five percent discount rate, in accordance with accepted benefit-cost analysis techniques.

Benefits

The benefits of the Protocol essentially represent the differences between two futures—what will likely occur with the Protocol in effect, and what would likely have occurred without the Protocol.

The global benefits consist of a series of health impacts plus quantifiable economic benefits associated with reduced UV-B damage to agriculture and forest resources and aquatic ecosystems, and to building materials such as polyvinyl chloride (PVC) plastics.

With respect to human health benefits, the study estimates that, as a result of the Montreal Protocol, and over the time period until 2060, there will be:
1) 19.1 million fewer (i.e. avoided) cases of non-melanoma skin cancer;
2) 1.5 million fewer cases of melanoma skin cancer;
3) 1/3 million fewer deaths from skin cancer; and
4) 129 million fewer cases of cataracts.

Avoided UV-B damage to agriculture, forest and fisheries resources is valued at $429 billion (1997 U.S. dollars). Avoided damage to materials is estimated to have a benefit of approximately $30 billion.

The most significant sensitivity analysis result was the scenario where implementation of the Protocol would have been delayed by twenty years (i.e. 2007). Such a delay would be expected to result in as many as 198,000 additional deaths, eight million additional cases of skin cancer and thirty-four million additional cases of cataracts.

Costs

The total estimated cost to global society of meeting the Protocol's control requirements is approximately $235 billion (1997 U.S. dollars, discounted over 1997–2060 time period). The shift from chlorofluorocarbons (CFCs) to alternatives, particularly in refrigeration and air conditioning applications accounts for the largest share, approximately $128 billion.

The estimated costs associated with conversion away from other ozone-depleting substances are:
1) methyl chloroform $47 billion;
2) hydrochlorofluorocarbons (HCFCs) $33.1 billion;
3) halons $12.6 billion;
4) methyl bromide $ 7.8 billion;
5) carbon tetrachloride $ 5.7 billion.

Looking back, it appears that the initial cost estimates were over-estimates, for a variety of reasons. Many of the technological developments were clearly unanticipated. Also, the costs of some alternatives were less than the ones they replaced, for example aerosols and solvent cleaning. The switch from CFCs to hydrocarbons as propellants in aerosols has cut material costs by as much as eighty percent, generating a cost saving of about $5.3 billion globally, over the period until 2060.

Adjustment costs were kept down because of a number of factors, including
1) international cooperation and technology transfer;
2) flexible implementation of the Protocol requirements by governments; and
3) industry cooperation in developing and testing substitutes for ODS.

Challenges

Not surprisingly, there were a number of "challenges" that had to be dealt with during the study. The most difficult ones were:
1) the most appropriate time frame over which to evaluate the benefits and costs;
2) the uncertainties regarding quantitative estimates; and
3) the evaluation of health impacts.

The time frame of the study was more controversial than first thought. The basic problem is that of trying to comprehend the validity of looking seventy-three years into the future.

Quantification on a global scale of the estimates of impacts of the Protocol was difficult to do. For many of the estimates, this study was the first attempt to do this. Although there have been a number of country-specific benefit and cost studies done over the years, they have not been done for all countries, particularly on the benefit side. To deal with these uncertainties, the approach taken in the study was to draw on the best available published evidence and supplement this with input from the International Advisory Panel.

The last "challenge" I want to mention concerns the valuation of health impacts. Many studies of individual countries have used various valuation approaches to convert health benefits into monetary terms. However, there are no widely accepted approaches available that could be applied on a global basis. In consultation with the International Advisory Panel, we decided that it was not feasible to develop valuation approaches for the wide range of effects (from cataracts to deaths from skin cancer) that could be applied on a global basis. Consequently, we limited the study to quantifying the health effect benefits, and did not try to come up with global monetary estimates for them.

Conclusions

The key conclusion of the study is that global society will realize significant net benefits as a result of implementation of the Montreal Protocol. The net benefits are estimated to be $224 billion, plus the health benefits described above. This demonstrates that the decisions taken were good ones, and justified economically as well as on health and environmental protection grounds. The sensitivity analysis on effects of delay in taking action demonstrates clearly the importance of moving quickly on this particular problem.

An interesting finding is that the costs of implementing the Protocol have been lower than initially expected. Recognizing it and appreciating the reasons why will be useful for other issues.

The reasons for the success of the Montreal Protocol are important to understand, and will serve us well in future environmental issues. The key reasons for the success are:
1) *Agreement on benefits.* From about the mid-eighties, major stakeholders were more or less in agreement on the science of ozone depletion and its effects. As the extent of the agreement increased, so did the controls of the Protocol.

2) The *science & technology basis* for decisions, and equally as importantly, the *process* set up to advise Parties (UNEP Assessment Panels).
3) The Protocol's *focus on targets rather than instruments* allowed countries the flexibility needed to meet the requirements in the most effective and efficient way for their particular circumstances.
4) *Accommodation of country differences.* The Protocol recognized and accommodated differences among countries in their ability to deal with the reduction and elimination of ozone-depleting substances—initially developing countries and subsequently, countries with economies in transition.
5) *Cooperation and technology transfer* have been the key to cost-effective global reductions. It has occurred at many levels: among governments and quasi-governmental organizations, private sector firms and associations.

In summary, this study has documented the overwhelming success of the Montreal Protocol in terms of benefits and beneficial impacts, in comparison with the costs of achieving those impacts. It provides clear evidence that a delay in introducing the Protocol would have led to significant widespread additional health and other impacts. Action at an early stage was possible because of the direct input of and response to scientific and technical information about the benefits and feasibility of controls. This illustrates the importance of integrating science and policy inputs in international issues like ozone depletion.

Notes

[1] The Protocol's real resource costs should not be confused with the costs of the Multilateral Fund established under the Protocol. The Fund represents only the cost to governments of funding specific initiatives to help developing countries in their transition away from the use of ozone-depleting substances

LESSONS FROM THE CFC PHASE-OUT IN THE UNITED STATES

Elizabeth Cook

How are efforts to protect Earth's ozone layer faring? Year after year, scientists find the seasonal Antarctic ozone hole is growing in size, depth, or duration and the ozone shielding other parts of the globe is "thinning". Meanwhile, the press reports, chlorofluorocarbon (CFC) prices are skyrocketing, trade in illegal CFCs is thriving, and attempts to roll back U.S. laws that safeguard the ozone layer—such as the 2001 ban on the pesticide methyl bromide—are as dogged as ever. Clearly, the job of halting ozone depletion is far from finished.

Yet, by key political measures, progress is undeniable. On January 1, 1996, the United States, the world's foremost consumer of ozone-depleting chemicals, met the internationally-set deadline for phasing out production of CFCs for domestic use.[1] Not that long ago, reaching this ambitious goal without devastating the economy was considered to be virtually impossible. Today the "elements of success" that made meeting the deadline possible can be documented and viewed as models not only for protecting the ozone layer, but also for tackling other tough environmental challenges, such as climate change.

To put these success stories into perspective, consider how great the challenge of phasing out CFCs seemed less than a decade ago. In 1987, the year governments signed the Montreal Protocol on Substances that Deplete the Ozone Layer, Americans were literally surrounded by CFCs: these "wonder chemicals" helped cool their cars and offices, preserve and package their food, and manufacture their new, high-tech computer and electronics gear (see Figure 1). U.S. industries were responsible for using one-third of all CFCs produced worldwide-and U.S. companies sold more than $500 million worth of the chemicals every year (Fay 1987, 180). American goods and services involving CFCs were worth $28 billion annually and installed equipment worth more than $128 billion relied on CFCs (Barnett 1987, 164). Not surprisingly, industry uniformly argued that moving away from CFCs would be a prohibitively costly, slow process that would harm the quality of products and services.

But the worst never happened. Today, industry has developed alternatives for virtually all CFC applications. The public has not been denied popular products, and the economy has not been seriously disrupted. In case after case, firms have eliminated CFCs faster, at lower cost, or with greater technological improvements than ever imagined. In fact, the whole phase-out is proving less costly than the Environmental Protection Agency originally estimated.

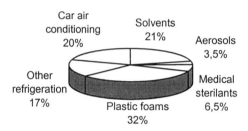

Figure 1. 1987 U.S. CFC End

In a 1996 study, the World Resources Institute and its collaborators explored a number of case studies to understand these achievements (Cook 1996).
1) Market forces played a major role in driving manufacturers of large building CFC-based chillers to overcome fears that the CFC phase-out would set back progress in making chillers more energy efficient. The industry once thought it would take eleven years to market air conditioning technology that could operate with new refrigerants—if, indeed, such a refrigerant could be developed. Yet, in less than six years, chiller-makers were marketing alternative systems that ran forty percent more efficiently than models built in the late 1960s and the 1970s—and twenty percent better than even some 1990 models.
2) The U.S. military, which helped develop fire-fighting halons in the 1940s and became their single largest user, provided the leadership necessary within the United Nations Montreal Process to end halon production in industrialized countries on January 1, 1994. Although halons were the least controlled ozone-depleter included in the 1987 Montreal Protocol, they eventually became the first to be phased out.
3) Extensive auto industry testing and CFC recycling have cushioned the economic impact of the phase-out on consumers who own the ninety-six million CFC-using cars still on the road. Early on, the auto industry assumed that retrofitting a car air-conditioning system to use CFC alternatives would be

complex and cost more than $1,000. Using much simpler options, service technicians can now satisfactorily retrofit most working car air conditioners for $50 to $150.

These examples of how the United States made the deadlines to phase out CFCs and halons do not tell the whole story, however. The U.S. government response to ozone depletion included an array of international, national, and local legislation as well as a range of regulations, bans, new product reviews, and economic instruments. How these actions influenced ozone-protection efforts is hard to measure since separating out cause and effect is difficult. Yet, a strong storyline does emerge from these successes in ozone protection, and five important lessons can be drawn.

1. An environmental goal that can be adjusted to reflect new scientific information is crucial

Without question, the main force motivating producers and users to invest in alternatives was the international community's decision to control CFCs and halons through the Montreal Protocol and the eventual adoption of the phase-out goal itself. The original 1987 Montreal Protocol required industrialized countries to halve CFC consumption by fifty percent by 1998 and freeze halon consumption in 1992 (United Nations 1987). Through adjustments in 1990 and 1992, Parties agreed to a full phase-out and pulled the dates forward to January 1994 for halons and January 1996 for CFCs (United Nations 1991 and 1993) (see Figure 2). The phase-out schedules—toughened in response to worsening ozone depletion—set the ground rules in producing and consuming industries and assured that the environmental goal would be met.

2. Economic instruments can help government and industry achieve environmental goals with greater flexibility and at lower cost

The United States did not take a traditional command-and-control approach to implementing the phase-out. It did not specify to each and every CFC- and halon-user precisely how to eliminate these chemicals. Instead of prescriptive regulation, the U.S. government used a combination of regulatory and market-based policies. Two innovative tools in this kit were the Environmental Protection Agency's (EPA) system of tradable consumption permits and Congress' tax on ozone-depleting chemicals.

The permit system that controlled the production and importation of CFCs and halons left users relatively free to decide how to meet reduction targets. The tax raised revenue and gave users a financial incentive to conserve the chemicals and adopt alternatives. Together, these market-based policies reduced administrative costs for EPA and lowered business' record-keeping expenses and compliance costs as firms searched for the least expensive compliance strategies. They also provided a regulatory structure that could respond quickly to new scientific developments. The results of a tightening cap on consumption and escalating tax on CFCs have been extraordinary: in the first four years that both policy tools were in force, CFC consumption was thirty-five percent below the allowable limit (see Figure 3).

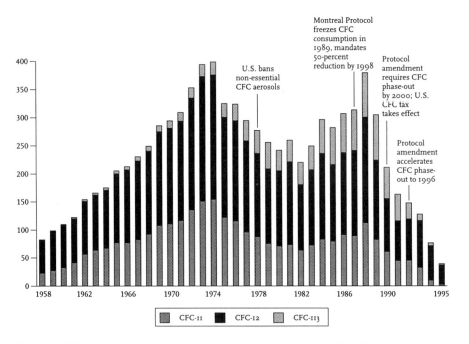

Sources: 1958–1993, U.S. International Trade Commission; 1994–1995, U.S. Environmental Protection Agency.

Figure 2. U.S. Production of Major CFCs and Policy Developments (thousand metric tons)

16-475 COCOA	E/M DEMONSTRATION ---------- DISCUSSES HOW QUANTITATI[VE ...] BE ACCOMPLISHED USING A
16-577 COCOA	FLAME PROPAGATION ---------- A DEMONSTRATION OF FLAME[...] DISCUSSION OF THE HAZARD[S ...] MISUSE OF EXPLOSION-PROO[F ...]
16-742 COCOA	SPECTROSCOPY ---------- INTRODUCES THE STUDENT [...] LIGHT IS PRODUCED WHEN [...] PASSED THROUGH GASES IN
16-743 COCOA	TRANSFORMERS ---------- DEFINES TRANSFORMER ACT[ION ...] TURNS RATIO, AND COEFFI[CIENT ...] ILLUSTRATES THE RELATIO[N ...] OF VOLTAGE AND CURRENT
16-745 COCOA	IDENTIFYING AND PRECUTT[ING ...] ---------- CHARACTERISTICS OF VCLA[...] SHIPS, HOW TO COMPUTE C[...] IDENTIFICATION

The permit system and tax are the cornerstones of U.S. ozone policy, but other policy decisions and the Clean Air Act Amendments of 1990 have also played a role. As mentioned, the firm phase-out date prompted many users to act earlier than required by law in anticipation of the eventual ban. Clean Air Act regulations mandating CFC recycling, product labeling, the development of safe alternatives, and the ban of nonessential uses have changed domestic consumption patterns of the controlled substances. Designed to inform the market and smooth the transition to alternatives, these steps have also required EPA to muster additional staff and resources.

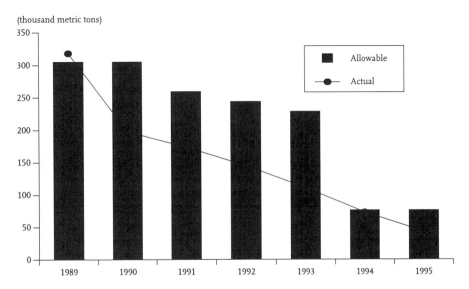

Note: CFC consumption appears above the allowable limit in 1989 because the controls did not take effect until July 1989.
Source: U.S. Environmental Protection Agency, unpublished data.

Figure 3. U.S. CFC Consumption: Allowable vs Actual

Is this combination of regulatory tools bringing about the CFC phase-out at the lowest possible cost? Nobody knows, but the transition to alternatives is costing less than EPA originally expected (see Table 1).

Table 1. Changes in Cost Estimates of CFC Regulations

Date of analysis	Policy	Social costs (1989–2000)	Cost/Kg
1988	50% by 1998	$2.7 billion	$3.55
1992	100% by 2000	$3.8 billion	$2.20
1993	100% by 1996	$6.4 billion	$2.45

Source: RIA (see text).

3. Innovative government initiatives can remove barriers that keep industry from solving environmental problems cost-effectively

Along with a market-based policy framework, government agencies have also taken an entrepreneurial approach to helping users find less costly CFC and halon alternatives. Once government redefined its adversarial relationship to the regulated community and decided to serve instead as a catalyst for change, agencies such as the EPA became information brokers for the private and public sectors—ready to help users scale barriers to adopting alternatives and to help entire industries phase out ozone-depleters cost-effectively.

For example, the EPA made collaborators of old adversaries in the search for solutions to ozone depletion. In 1988, the agency invited industry and environmental groups to participate in negotiations that led to a voluntary CFC phase-out by the food packaging industry in less than a year. Theses companies were subsequently allowed to use hydrochlorofluorocarbon-22 (HCFC-22) as a foam-blowing agent on an interim basis, which enabled them to quickly convert a plant for $50,000 instead of spending up to $2 million to adopt hydrocarbon-based technology. Since air-pollution controls on hydrocarbons were costly and market pressures to rapidly eliminate CFCs strong, the option to use HCFC-22 allowed manufacturers to protect the environment without closing factories or laying off workers.

EPA also initiated the formation of the Industry Cooperative for Ozone Layer Protection (ICOLP), which fostered cooperation rather than competition in the search for CFC solvent alternatives. In this forum, major electronics firms together perfected "no clean" technology—an underdeveloped Canadian process for which the patent had expired—and essentially re-engineered the soldering process to eliminate the need for CFCs. The result of this unprecedented technical collaboration: a more efficient and effective technique. One company that achieved a phase-out in three years, Nortel (formerly Northern Telecom), estimates that it invested $1 million to purchase and employ new hardware and saved $4 million in chemical waste-disposal costs and CFC purchases.

"No clean" technology caught on partly because the Department of Defense (DoD) was also willing to collaborate and bend. Military specifications, which required electronics manufacturers to use CFC-113, kept firms wedded to CFC use. EPA and DoD jointly created the Ad Hoc Solvents Working Group to work with industry to remove this roadblock to the phase-out. Through this partnership the main culprit "milspec" (Military Standard 2000) was revised in record time, allowing for the use of "no clean" technology and other alternatives that could pass a benchmark test of cleaning "as good or better" than CFC-113. The shift served as a model for a broader transition from prescriptive milspecs to performance-based standards within DoD.

4. Given the opportunity, industry leaders can find ways to innovate and gain competitive advantages in response to environmental challenges

The need to eliminate CFCs has forced firms to redesign products and operations. True, substantial human ingenuity and capital investment were required, but many companies that made meeting the CFC phase-out a priority transformed a potentially costly liability into a competitive advantage. Winners have boosted product performance, found new ways to be efficient, developed low-cost alternatives, and even saved money. Chiller efficiency improvements and "no clean" technology hammer this point home.

A coalition of utilities, environmental organizations, and government officials tapped into industry's innovative and competitive spirit by creating a $30-million "Golden Carrot" prize to encourage companies to develop a super-efficient refrigerator/freezer as they redesigned their appliances to eliminate CFCs. Whirlpool won the competition, and in 1994, introduced a new model that exceeds the Department of Energy (DoE) efficiency standards by almost thirty percent. Financial incentives and market opportunities helped manufacturers overcome their initial view that they could not meet DoE standards and eliminate CFCs too. What is more, the incentive pulled an appliance with advanced technologies onto the market in less than two years, instead of the eight or nine typically needed.

5. Pre-regulatory cost estimates often far exceed the actual cost of complying with environmental regulations because they fail to reflect the technological innovation environmental policies spark

Because no one could foresee all of the possible solutions and innovations in advance, early estimates of CFC control costs often turned out to be higher than actual costs. The U.S. phase-out of CFCs illustrates the limits of economic models that aim to estimate the costs of regulations but take a static view of the world. Important influences on compliance costs—how firms respond to environmental goals, flexible market-based incentives, and innovative government initiatives—often fall outside the compass of these assessments.

Before issuing CFC regulations, the U.S. EPA funded economic evaluations of the range of possible control options for reducing emissions and of the associated social costs. In the early 1980s, the studies examined selected CFC-using sectors

(Mooz et al. 1982; Palmer et al. 1980). More recently, through Regulatory Impact Analyses (RIAs), EPA also calculated a total cost for implementing the Montreal Protocol and the Clean Air Act in the United States. EPA updated these analyses as governments adjusted and amended the Protocol, tightening the controls on ozone-depleters.

Because EPA changed the scope of each economic analysis to reflect the different policy options under consideration, it is difficult to find a consistent measure of CFC control costs. But, in general, the revised RIAs illustrate that the economic models initially overstated regulatory compliance costs. Consider the estimated social costs of CFC controls over the period 1989 to 2000. In 1988, RIA experts estimated that it would cost $2.7 billion to halve U.S. CFC consumption within ten years—an average cost of $3.55 per kilogram reduced (Lee 1988). Four years later, the RIA estimate for a total CFC phase-out by 2000 was $3.8 billion. Even though eliminating the remaining fifty percent of CFCs would be harder than getting rid of the first half, the estimated average cost had dropped thirty-eight percent to $2.20 per kilogram (Lee 1992). Finally, in 1993, yet another RIA evaluation of an accelerated CFC phase-out estimated that the January 1, 1996 phase-out-would cost $6.4 billion. The cost of deeper and faster CFC reductions—and additional environmental benefits—raised the estimate of per kilogram costs only slightly to $2.45 (See Table 1.).

The economists who developed the 1987 RIA did not, of course, deliberately overestimate the costs of CFC reductions. Rather, they used the best information they had on available control options. These examples cover only a handful of the unexpected technological advances that had to be factored into analysts' subsequent cost estimates. Even the DuPont Company—the world's largest manufacturer of CFCs and perhaps best positioned to predict how CFCs users would respond—substantially revised its projection of the market for its chemical substitutes. In 1989, DuPont estimated that its HCFC and hydrofluorocarbon (HFCproducts would capture thirty-nine percent of the CFC replacement market, while roughly one third of the market would be lost to non-fluorocarbon alternatives. By 1992, DuPont was pegging the market for its substitutes at only twenty-six percent and expected non-fluorocarbon solutions to capture almost half the market (DuPont) (see Figure 4).

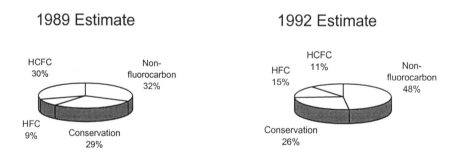

Figure 4. Changes in DuPont's Estimates of How CFCs Might Be Replaced

Teasing out the various forces that drove ozone-protecting innovations reveals many common themes, but it also illustrates why economic models could not possibly incorporate the complex events that led to each answer. Once CFC-users began an earnest search for alternatives—within a policy framework that stimulated innovation—the estimated cost of compliance fell rapidly.

In addition to these five lessons, it is important to note that these responses were not sheer acts of will. In each case, government and industry were moved to action by scientific developments, intense media coverage, rising public awareness, and effective environmental advocacy. Corporate leadership also became a powerful force for controlling ozone-depleting chemicals. Companies that took bold early steps to eliminate CFCs dramatically influenced their competitors, suppliers, and employees, as well as policy-makers.

Future Challenges

Environmental and industry leaders continue to call attention to the steps still needed to put the use of ozone-depleting chemicals behind us. Innovative approaches and further cooperation are needed to meet these future challenges. Can the military sparingly use its halon "reserve" and help develop safe technology eventually to destroy these potent ozone-depleters? Will the cost of retrofitting vehicular air conditioners continue to decline without the proliferation of new refrigerants in the marketplace causing a cross-contamination problem and damaging systems? Is the crackdown on CFC smugglers strong enough to destroy the black market? And, perhaps, above all, will science and industry continue to search for CFC alternatives that do not harm the ozone layer and that do not add to global warming? Many of the substitutes that industry has developed to replace CFCs still deplete ozone, though less potently, and HFC-134a—the widely adopted refrigerant—is itself a greenhouse gas. The super-efficient Golden Carrot refrigerator, most new chillers, and new or retrofitted car air conditioners all rely on these compounds.

As scientists develop ever more information about the impact industrial society is having on Earth's atmosphere and as future responses are fashioned, the lessons of the CFC phase-out should be kept in mind: set firm goals; accord industry flexibility in meeting them; make use of economic incentives; and allow government to serve as a catalyst for change. If history is a guide, policies that embody these elements will stimulate innovative solutions.

Tackling Climate Change

The importance of firm environmental goals is nowhere more clearly seen than in current efforts to avert climate change. The experience of the United States and many other developed countries in trying to meet greenhouse gas reduction pledges in the Framework Convention on Climate Change has been disappointing mainly because there has been no enforceable target. Good intentions aside, national commitments to reduce greenhouse gases cannot be met by voluntary industry actions alone. A strong signal—an emissions-reduction target—is required to spark needed change and investment.

Similarly, how can government determine in advance the best way to lower emissions from each source of a greenhouse gas? Marketable carbon permits, carbon taxes, or other market-based approaches would allow each firm to respond to a reduction target in a way that costs the least. Besides wielding such economic instruments, government should also help remove or reduce barriers to investments in climate-friendly technologies and provide information on using such technologies cost-effectively to cut emissions. In one example, the U.S. Environmental Protection Agency's Green Lights Program, government encourages businesses and organizations to adopt energy-efficient lighting: participation is voluntary and government's role is a far cry from that of regulator. In another, government can procure greenhouse gas reduction technologies—whether high-efficiency chillers, renewable energy systems, or electric vehicles.

On a positive note, emerging U.S. policy on climate change suggests that policy-makers are learning these lessons. In July 1996, at the Second Conference of the Parties to the Framework Convention, the United States admitted that the treaty's non-binding structure has failed and called for legally binding international greenhouse gas emissions targets, the use of market-based policies, and the establishment of a long-term greenhouse gas concentration goal to guide private investment (Wirth 1996).

What should such targets and goals look like? The government has not said, but as debate heats up over the specifics, policy-makers and companies unmoved by the many CFC success stories claim that binding greenhouse gas emissions targets will be difficult to meet and economically disastrous. Continue scientific and technological research they say, but wait to take further action since climate-protecting technologies and mitigation measures will cost less in the future. Such foot-dragging simply does not square with past experiences or with the underlying finding that smart policies will drive technology innovation and bring compliance costs down.

Notes

[1] January 1, 1996 is also the phase-out deadline for methyl chloroform and carbon tetrachloride. Under the Protocol, metered dose inhalers, rocket motor manufacturing, and laboratory applications have been approved as essential uses for 1996 and 1997. Also, U.S. companies can continue CFC production beyond January 1, 1996 for export to developing countries.

References

Cook, Elizabeth (ed.). 1996. *Ozone Protection in the United States: Elements of Success.* Washington D.C.: World Resources Institute.

DuPont Company. Changes in Estimates of How CFCs Might be Replaced. Handout.

Global Climate Coalition. 1996. "U.S. Business Faults Clinton Administration Climate Change Policy, Foresees Significant Negative Job, Trade, Economic Impacts." [News Release] Washington, D.C. July 17.

Lee, David [U.S. Environmental Protection Agency]. 1988. "Calculations Based on Regulatory Impact Analysis: Protection of Stratospheric Ozone." (Washington, D.C.: ICF Incorporated for EPA, August 1, 1988), vol. 2.

Lee, David [U.S. Environmental Protection Agency]. 1992. "Calculations Based on Regulatory Impact Analysis: Protection of Stratospheric Ozone. (Washington, D.C.: ICF Incorporated for EPA, June 4, 1992).

Lee, David. [U.S. Environmental Protection Agency] 1993. "Calculations Based on Addendum to Regulatory Impact Analysis: Protection of Stratospheric Ozone." (Washington, D.C.: ICF Incorporated for EPA, December 1993).

Mooz, W.E. et al. 1982. *Technical Options for Reducing Chlorofluorocarbon Emissions.* Santa Monica: The Rand Corporation.

Palmer, Adele R. et al. 1980. *Economic Implications of Regulating Chlorofluorocarbon Emissions From Nonaerosol Applications.* Santa Monica: The Rand Corporation.

Fay, Kevin J. 1987. "Statement." Prepared for the House of Representatives Subcommittee on Health and the Environment, Ozone Layer Depletion [hearings, March 9]. 100th Congress, First session.

Barnett, Richard. 1987. "Statement." Prepared for the U.S. House of Representatives Subcommittee on Health and the Environment, Ozone Layer Depletion [hearings, March 9]. 100th Congress, First session.

United Nations. 1987. *Montreal Protocol on Substances that Deplete the Ozone Layer.* 26 I.L.M. 1541 (January 1, 1989).

United Nations. 1991. *London Amendment to the Montreal Protocol.* 30 I.L.M. 537.

United Nations. 1993. *Copenhagen Amendment to the Montreal Protocol.* 32 I.L.M. 874.

Wirth, Timothy E. 1996. "Statement on behalf of the United States to the Second Conference of the Parties at Framework Convention on Climate Change." Geneva, Switzerland. July 17.

HIGHLIGHTS OF OZONE PROTECTION LEADERSHIP BY INDUSTRY IN DEVELOPING COUNTRIES

Jorge Corona and José I. Pons

Members of the Technology and Economic Assessment Panel (TEAP) believe it is important to recognize the contributions made by industry to protect the ozone layer, in both developed and developing (Article 5(1)) countries. However, it is in Article 5(1) countries that many multinational corporations have made some of the most significant impacts through voluntary initiatives. Despite the ten-year grace period allowed in Article 5(1) countries, many corporations have voluntarily upgraded their production facilities in these countries to the same standards as their developed country facilities, and avoided the temptation to dump obsolete industrial technologies dependent on ozone-depleting substances (ODS) in Article 5(1) countries.

While Article 5(1) countries are often grouped together in a way that does not distinguish one from another, they are in fact quite heterogeneous. They differ greatly in level of industrialization, climate, culture, and political and economic system. In addition, there is a broad range of ODS applications across a wide spectrum of industrial sectors, many of which are country specific. This leads to a wide range of ODS use patterns which are driven by parameters such as the number and size of subsidiaries of multinational firms operating within a country, the number of domestic small and medium-sized industrial users, as well as the availability of local ODS suppliers. As a result, ODS consumption and production patterns in Article 5(1) countries are extremely diverse.

As in developed countries, changes in industrial practices in Article 5(1) countries are driven by uncertainty. Companies are very reluctant to change, particularly when the direction and level of change are not well defined; change represents a risk to their business, their employees, and to shareholders. But change is what companies face when they confront the need to select an appropriate ODS alternative technology. In some countries and for some applications, the debate has not yet been settled regarding which is the most appropriate alternative. Moreover, when resources are limited, there is fear to invest in a technology that might not work, or will soon be rendered obsolete by a new and emerging alternative. So, there tends to be a perception that waiting is better because new technologies will be better proven over time. It is not unlike someone faced with the purchase of a new computer: the longer you can wait, the better machine you will get at a lower price; but you cannot wait forever, so how do you decide exactly when to make the purchase decision? Willingness to change is also viewed differently by different generations. Young industrial entrepreneurs are usually more willing to embrace change and new technology than those who have long been familiar with existing technology.

In many Article 5(1) countries, ODS use is so small that it becomes possible to define strategies for development that circumvent their use. This is an extremely sound and forward-looking strategy for a country to pursue. However, it requires political resolve and the cooperation of multinational corporations from developed countries. Vietnam provides an interesting case study. The Japan Electrical Manufacturers' Association (JEMA), the Japan Industrial Conference for Ozone Layer Protection (JICOLP), the International Cooperative for Environmental Leadership (ICEL—formerly the International Cooperative for Ozone Layer Protection, or ICOLP), the Japan Ministry of International Trade and Industry (MITI), and the U.S. Environmental Protection Agency (EPA) organized the Vietnam-U.S.-Japan Technical Meeting on Protecting the Ozone Layer. The government of Vietnam was key to the success of the project and committed to development that did not depend on ODS. As a result of this cooperative project, more than forty multinational companies from seven countries pledged to help the government of Vietnam protect the ozone layer by investing only in modern, environmentally acceptable projects in that country. This meant that companies operating in Vietnam would use the same ODS-free industrial technologies they used in developed countries. This enables Vietnam to leap over a type of development based on ODS, thereby avoiding a future industrial conversion. While conversion to non-ODS technologies may be partially funded by the Multilateral Fund, in reality, the conversion itself is a distraction to industry and a burden to development. This same approach could be adopted in other Article 5(1) countries wishing to develop efficiently by doing it right the first time.

Despite the virtues of these arguments in favor of moving quickly, there are always good reasons to resist change. Industry used their best arguments to resist the Montreal Protocol in the early and mid 1980s—questioning the science, and forecasting dire economic consequences if the proposed changes went forward, due to lack of cost effective alternatives. Promoters of methyl bromide are resurrecting these same arguments today. However, in addition to the arguments above, others are unique to Article 5(1) countries. In some cases, ODS or equipment that uses ODS, are dumped at bargain prices because there is no market in developed countries. This appears attractive in the short term. However, as Vietnam recognized, it may not be such a good deal in the long run. In other cases, societies and their economies may be isolated and sufficiently large as to be self-sufficient, therefore lacking incentives to phase out ODS. Other issues include the administrative burden required to obtain funding which may be suffocating for small enterprises, and in some cases patent or "intellectual property" issues that may limit access to new technologies and chemicals.

However, some important lessons have been learnt after ten years of the Montreal Protocol. The importance of identifying innovators and motivators is now well understood. These individuals and organizations are found in governments, non-governmental organizations (NGOs), and industries. They are the source of the solutions, but cannot be predicted or integrated into economic models. Leadership is also a crucial but unpredictable element. When political or corporate leaders set the direction, and when standards of industrial behavior require ODS free technologies, then the rest of the "pack" is forced to follow.

In addition to environmental reasons, there are many business reasons for Article 5(1) countries to phase out in advance of the required timeline. In some cases companies can realize enormous savings by moving to alternatives. For example, in the areas of aerosols, foams and some refrigerants, hydrocarbons are much cheaper than chlorofluorocarbons (CFCs). In many countries, better educated and more sophisticated urban populations are well aware of ozone layer depletion, and favor green alternatives for their industries. Companies which see future growth in exports to developed countries have no choice but to be ODS free—there is no market for ODS-based technologies or products. In addition, as the Montreal Protocol controls become applicable to Article 5(1) countries, those companies with ODS-free products and processes will enjoy competitive advantage over other countries and companies which chose to delay conversion.

In addition, there is funding available through the Montreal Protocol Multilateral Fund for the incremental cost of converting to ODS-free processes in Article 5(1) countries. To date, twenty thousand tons of ODS have been phased out thanks to Protocol funded projects, and there are projects to phase out an additional eighty thousand tons waiting to be implemented.

Finally, the health reasons for phasing out ODS may be more meaningful to Article 5(1) countries than elsewhere due to less developed health care systems. Many people who have experience with skin cancer or cataracts are particularly sensitive to the importance of protecting the ozone layer and less inclined to wait until the phase-out is mandated by the Protocol.

Multinational corporations are particularly well positioned to help Article 5(1) countries protect the ozone layer, both because of the location, size, and sophistication of their industrial facilities, and because many multinationals participate in the establishment of worldwide industrial and environmental standards. An example of early corporate leadership is Johnson Wax, the first U.S. company to have switched to hydrocarbon aerosols in advance of 1978 U.S. regulations controlling CFCs in aerosols. British Petroleum, Lufthansa, and Northern Telecom are also examples of early industry leadership in the Montreal Protocol. In many cases, concerned companies joined efforts to help others within their respective industries, either in their own countries or abroad. JICOLP and

ICEL are two good examples. Sometimes, however, efforts to achieve an early phase-out were not as successful. Technology co-operation requires effort to orchestrate information-sharing, to motivate other industries and customers, and to secure financing either from within a corporation, from a government agency, or from the Multilateral Fund. In addition, industry leadership is not limited to firms from developed countries. As evidenced by successes in Vietnam, companies from Article 5(1) countries have also demonstrated leadership.

One of the achievements of the Montreal Protocol has been the creation of individual country programmes and of Ozone Offices in virtually every Article 5(1) country. Information about local ODS use and appropriate substitute technologies is available through the United Nations Environment Programme (UNEP) as well as through technology suppliers, clients, and trade associations. Now that substitute technologies are widely available and resources can be obtained through the Montreal Protocol, it becomes evident that user motivation is vital to complete the phase-out process. With the first control measures for Article 5(1) countries effective in 1999, there are likely to be compliance problems if action is delayed. Industry needs time to prepare projects, identify substitutes, and obtain financing. In addition, Protocol and Multilateral Fund resources are limited and some projects may have to wait until funding is available. Further, if economies continue at their current growth rates, increased emissions from Article 5(1) countries will negate the benefits achieved to date by developed countries. As Table 1 shows, ODS production and consumption in Article 5(1) countries are up significantly since the Montreal Protocol baseline year.

Table 1. 1995 ODS Production and Consumption Data For Article 5(1) Countries (in ODP metric tonnes/year)

Substance	PRODUCTION		CONSUMPTION	
	1995	Base	1995	Base
CFCs	107 295	38 764	156 599	109 046
Halons	40 750	10 600	35 295	33 390
Met.Brom.	192	74	6 168	3 105
TOTAL	148 237	49 438	198 062	145 541

This point is clearly illustrated by the official numbers UNEP has gathered from Article 5(1) Parties. The numbers show production and consumption data for 1995 and the respective base year for CFCs, halons and methyl bromide in ODP tons. The Methyl Bromide Technical Options Committee believes the Methyl Bromide numbers may be underestimated.

Significant trends can be inferred from these figures. Despite many examples of accelerated phase-out, Article 5(1) Parties' consumption of ODS has increased thirty-six percent from base year data, and production increased two hundred percent. These figures represent only seventy-five percent of local consumption. These trends are an important reminder that the effort to protect the ozone layer is not over yet, and significant challenges remain in Article 5(1) countries.

COMPETITIVE ADVANTAGE THROUGH CORPORATE ENVIRONMENTAL LEADERSHIP

Margaret G. Kerr

This paper presents a topic about which I feel strongly—the business benefits of environmental leadership. Some of you might be confused by my title, Senior Vice-President for Employee and Customer Value, because if has no mention of environment. But I strongly believe that Nortel's environmental activity is a key contributor to employee and customer value, my portfolio at Nortel.

Since 1987, I have been head of environmental affairs at Nortel. My portfolio at Nortel has expanded in recent years, and now includes human resources, employee satisfaction, health and well being, safety, security, business ethics, and customer satisfaction, in addition to environmental affairs. The exciting thing for all of us in the function has been coming to a fuller understanding of the close relationships among these seemingly disparate activities—how they can work together in synergy to support the business objectives of the corporation.

By way of introduction, Nortel, is the leading supplier of digital telecommunications networks in the world. We are a global company, now in our second century of business and we are headquartered in Canada. Our employees—about sixty-eight thousand strong—operate in more than 100 countries. Today, nearly forty percent of our business comes from outside North America. Last year our revenues were U.S.$ 12.8 billion.

At Nortel, we believe that minimizing the impact of our products and operations on the environment is part of our ethical responsibility as a global corporate citizen. Clearly, we owe it to the communities in which we do business around the world to comply with local legislation and guard against environmental accidents. Wherever we can, we also want to serve as a positive force, actively contributing to environmental improvement.

We have learned over the years, however, that environmental protection is not just the "right thing to do"—it actually translates into competitive advantage for the company. As you will hear shortly, advantage has been gained in areas such as lower operating costs, product and service differentiation, and improved corporate image. Most importantly, environmental protection can create customer value.

I will not attempt to relate the rich examples of competitive advantage gained through environmental protection by other companies—they would do a much better job. Instead, I will concentrate on Nortel's experience, not because I think we are unique in the benefits we have gained through environmental protection, but because it is what I know best.

The evolution of environmental management at Nortel

I think the evolution at Nortel accurately reflects the changes in thinking that have occurred in industry as a whole. Our understanding has both broadened and deepened as we have come to see environmental management as an integral part of the business of the company—not as a discrete activity that sits off on the side somewhere.

I think our approach to environmental management has been influenced—sometimes consciously, sometimes less consciously—by the evolution of quality management. The history of the quality movement began with the recognition that "quality control" did not have to mean the costly practice of weeding out and scrapping defective products before they were shipped to customers. "Doing things right the first time" could lead to better products at lower costs. Companies started considering the manufacturing process as a system, and looking for opportunities to reduce errors, scrap, rework—and costs.

In recent years, notions of quality have been migrating to a broader concern with customer value management. Companies are focusing on getting close to the customer, and being driven by the customer's needs and expectations. They are also trying to get closer to the market—learning how their product quality, service quality and price stack up relative to the competition. They are moving from a narrow, internal focus to an ever-widening understanding of the broader context—the larger systems of the business's operations. They are thinking about capturing a larger share of the customer, as well as a larger share of the market—creating a base of loyal customers.

There has been a progression in thinking about environmental management that has some rough parallels with this evolution. It is a progression first from after-the-fact control to upstream prevention through a systematic management approach based on the principle of continuous improvement. It then broadens to an understanding of how environmental activity can add value to customers, and differentiate a company from its competitors in the marketplace.

It is no secret that for a good part of the 1980s, most of those in industry who thought about environmental issues at all tended to have an image of costly problems that would have to be "fixed" when they became too pressing. "Environmental management" was really pollution control—putting scrubbers on smokestacks, treating polluted wastewater, and ensuring that hazardous waste was managed as safely as possible. These activities were a bottom-line cost to the company—things that industry chose to do in response to public pressure or was forced to do by increasingly stringent government regulations—but very definitely detrimental to the company's profitability.

At some point in the 1980s, we began to realize that we really should be thinking about pollution prevention rather than pollution control. And in so doing, we imported an idea from quality management: the idea that "prevention" required a more sophisticated understanding of systems. We needed to look at our manufacturing processes as a system and figure out where there was waste, inefficiency or unnecessary pollution.

This was brought home to us in the late 1980s/early 1990s by our chlorofluorocarbon (CFC) elimination project. We started out thinking that our challenge was to find a less harmful alternative to the CFC-113 solvent we used to remove flux residue from printed circuit boards. But instead of switching to a different solvent, we ended up redesigning our technology and processes to eliminate the need for cleaning altogether. And we showed that environmental protection could actually save money.

In addition to eliminating the use of this ozone-depleting substance (ODS) years ahead of the deadline set by the Montreal Protocol we managed to realize a pretty good return on investment that considerably heightened senior management's interest in pursuing environmental initiatives. We spent $1 million on research and development, but in the three-year duration of our "Free in Three" project alone, we saved about $4 million. The savings came from decreased solvent purchases, the elimination of cleaners and their associated operating and maintenance costs, and reductions in solvent waste for disposal.

Through our work with engineering and manufacturing personnel on the elimination project, our environmental staff was exposed to the current thinking about quality management systems. They started to see that we had been thinking about environmental management more as a checklist of tasks to be accomplished than as an ongoing system of continuous improvement. We realized that we needed to develop a "systems view" of the way we managed our environmental activities.

Our revised environmental policy stresses the importance of a systematic approach, and institutionalizes our corporate Environmental Management System—or EMS—Standard. Finalized and piloted in 1994, the EMS Standard is now being adopted by manufacturing and research locations throughout the company. Based on the principles of quality management and continual improvement, our EMS Standard was designed to reflect the requirements of ISO 14001. In fact six of our sites in Europe have now achieved certification to ISO 14001.

Our experience has been that the implementation of an EMS contributes to competitive advantage in two ways. First, it is making us take a hard look at all of our processes and root out costly inefficiency and waste. Second, as more of our customers include evidence of a systematic approach to environmental management in their supplier selection criteria, our implementation of the EMS Standard shows that we are listening and responding. British Telecom, for example, keeps a list of suppliers with environmental management systems that meet their standards—Nortel was one of the first companies to make the grade.

While the EMS standard sets out a systematic framework for managing environmental impact, it does not prescribe what to do. In the initial stages, we focused on setting and achieving measurable targets for reducing the environmental impacts of our manufacturing processes. Emerging ideas of product stewardship or product lifecycle management have now made us more conscious of the environmental impact of our products themselves. While we have been incorporating product stewardship ideas into our thinking over the past several years, in 1995 we increased our emphasis on minimizing the environmental impacts of our products throughout their life cycle—from design through to the end of their useful lives.

In developing our product lifecycle management initiative, I think we are more driven by the needs and expectations of our customers than we have been before. By maximizing the environmental and economic efficiency of our products at each stage in their life cycle, we are adding value for our customers and reducing costs for Nortel. I shall present only three examples of what we have worked on to date in the product design, manufacturing, and distribution stages:

1) Interventions in the product architecture stage—the very first stage of product design—can have a major impact on the resource efficiency of a product. The designers of our Nortel PowerTouch telephone have taken a modular approach to design—one that allows customers to upgrade their telephones without buying new ones and scrapping the old. The PowerTouch consists of a standard base, and an upgradeable slide-in module that can add features such as incoming caller ID, call waiting, or a larger screen and better graphic display. The module can be replaced over time to provide the latest features available, at half the price of replacing the phone. This also reduces the volume of obsolete product sent for recycling or to landfill.

2) A change in the way we think about supply management is proving the value of looking for opportunities to reduce environmental impact and cost in all parts of the value chain. We have recently launched a pilot project with our main chemical supplier at a Nortel site in Ottawa under which we are purchasing the supplier's services for a fixed fee, rather than purchasing the chemicals themselves. In addition to the cost of the necessary chemicals, the fee covers services such as chemical process expertise, chemical management, storage, and disposal. Under traditional supply arrangements, it is in the supplier's interest to encourage Nortel to use more chemicals. The new arrangement provides an incentive to the supplier to help Nortel minimize chemical use, to develop more efficient chemical processes, and to find alternatives to hazardous chemicals.

3) Minimizing the cost and environmental impact of product packaging is a key concern of our customers. Packaging innovations such as collapsible carts, aluminum carriers, plastic "clamshell" packages and standardized pallets and boxes—all reusable—are helping us reduce costs and packaging volume. A single change in our distribution practices is saving an estimated $5 million

annually: we are now assembling switching products before they are shipped, rather than packaging and shipping components separately for on-site assembly. What is known as "plugs in place" shipping not only requires less packaging, but also allows for faster installation.

So far I have addressed how our thinking and action around environmental protection have evolved from end-of-pipe control to a management systems approach that is focused on prevention and based on quality management principles. I have also addressed how our new thrust in product lifecycle management is leading us to become more customer-driven. I now want to take a broader look at what we have learned about how environmental performance contributes to competitive advantage by enhancing a company's image.

The CFC story did not really end when we became, in January 1992, the first multinational company in the electronics industry to eliminate CFC-113 from operations worldwide. In subsequent months, we received several prestigious environmental awards, and got a lot of positive media attention—which when you get right down to it is free advertising for the company. While we think of it as old news, we are still getting requests to write articles, give speeches, or be the subject of television spots on CFC elimination.

More importantly, we have been deeply engaged in a process of sharing what we have learned with other countries—especially "developing" countries. Between 1992 and 1995, Nortel played a lead role in technology cooperation projects in Mexico, India, China, Turkey, and Vietnam. These projects were supported by World Bank funding, and involved close collaboration with local government and our partners in the International Cooperative for Ozone Layer Protection (now the International Cooperative for Environmental Leadership).

Over the years, we have come to believe strongly that international cooperation between governments and industry is a highly practical way of resolving shared environmental problems. In part, we have been devoting substantial amounts of time and energy to this because we believe that it is part of our responsibility as a global corporate citizen. But our reasons are not just altruistic. Technology cooperation is a marketing tool: it is building goodwill and strong relationships with customers in emerging markets. Our willingness to bring first-tier technology and our experience with environmental management to developing countries—helping them avoid some of the costly mistakes we have made—is a key driver in many of our joint ventures. Our commitment to the environment is a part of the value proposition we offer to potential customers.

Increasingly, our willingness to work with customers to solve shared environmental problems is helping us build customer satisfaction and loyalty in developed markets as well. For example, for years we have been taking back obsolete equipment from Bell Canada and routing it through a Nortel material recycling facility. We actually have three such facilities in North America and one in the U.K.—all of them revenue generators. We are now finding that companies such as BT and Mercury Communications in the U.K., as well as Telia of Sweden, are actively seeking suppliers who will take on the responsibility of disposing the equipment they sell at the end of its useful life. Several European countries already have product take-back legislation (also called "producer responsibility" legislation) in place, and the European Union is considering enacting uniform standards. We are now putting our experience with product recovery to good use in highly successful programs with BT and Mercury. Our commitment to the environment is helping us build strong and, we hope, long-lasting relationships with these key customers. Product recovery is an excellent example of how we can provide new customer services while reducing the environmental impact of our products.

I have talked a lot today about Nortel—not because I think that what we are doing is so far in advance of what other companies are doing, but because I want you to understand how we have gradually come to see that environmental leadership can help Nortel provide superior customer value. Environmental programs that are put in place because they are the "right thing to do" or because governments require them are vulnerable. They are subject to the whim of legislators, swaying public priorities, and financial cycles. For long-term sustainability and impact, environmental activities must be seen by decision-makers at all levels of the organization to be clearly supportive of business objectives and a contributor to competitive advantage in the marketplace.

We are certainly a long way away from having environmental considerations take a front seat whenever a decision is made or an activity undertaken within Nortel. But that is our long-term vision. Critical to our success will be our ability to communicate how environmental management contributes to competitive advantage—and so once again I welcome today's opportunity to voice my thoughts on the subject.

LESSONS FROM THE THAILAND LEADERSHIP INITIATIVE, THE VIETNAM CORPORATE PLEDGE, AND THE GLOBAL SEMICONDUCTOR AGREEMENT

Yuichi Fujimoto

The Phase-out of Chlorofluorocarbons (CFCs) from Household Refrigerators in Thailand

At a seminar conducted for the Association of South-East Asian Nations (ASEAN) countries held in Singapore in 1990, it was reported that CFC consumption in Thailand was increasing while CFC consumption in Singapore decreased after reaching maximum consumption in 1988, and consumption in Malaysia also decreased after 1990. Only CFC consumption in Thailand was increasing—reaching 7,300 tons in 1990—and was estimated to exceed 10,000 tons in 1991. Enterprises from Japan and the U.S.A. operating in Thailand were responsible for the increase.

To address this problem, the first Japan-U.S.-Thailand Technical Seminar on Ozone Layer Protection was held in March 1992, in Bangkok, Thailand. It included participants from the Japanese, American, and Thai governments and from the industries using CFCs in Thailand. Experts in industrial cleaning, refrigerants and foams from Japan, the United States, and the European Union, as well as representatives from the Montreal Fund and the World Bank attended the meeting. Thirty Japanese and eleven American companies announced their own phase-out schedules for CFCs and 1,1,1 trichloroethane. At the meeting, Thailand announced its target date for eliminating CFCs from Thai household refrigerators—the end of 1996, only one year later than in Japan and other developed countries.

Seven Japanese companies produce household refrigerators in Thailand.

While the phase-out schedule was comparable to that of Japan, significant problems needed to be overcome before the schedule could be achieved. Compressors for domestic refrigerators were supplied by a single Thai company to four of the seven Japanese refrigerator companies in Thailand. As a result, it was crucial that the Thai compressor manufacturer redesign their products to use CFC alternatives and manufacture reliable compressors. As it turned out, an unprecedented degree of cooperation would be necessary to make this conversion successful. The Thai compressors had to be competitive with those from foreign suppliers, and they had to use the alternative refrigerant HFC-134a. The local compressor manufacturing company was scheduled to supply the new HFC-134a compressors in time for the new refrigerators to be produced by the end of 1996. Trial compressors were tested by the Japanese companies and they found the compressor parts and capillary tube to be contaminated with manufacturing residue. This caused the pipe in the refrigerant circuit to clog. There were additional reliability problems caused by poor lubrication and unacceptable wear. These problems made the compressors unreliable. Despite significant effort to resolve the problems, the local company found itself unable to produce reliable compressors in time for the phase-out.

As a result, the four Japanese companies waiting for the compressors from the local company had two options. Either keep to the phase-out schedule by getting compressors from manufacturers other than the local company or postpone the phase-out and wait for a supply of reliable compressors from the local company. Neither of these options was acceptable.

The Thai government was eager to keep the phase-out on schedule.

No other local compressor suppliers could be found with the capability to supply enough compressors to meet the deadline established by the Thai government. The Japanese companies nevertheless decided to keep to the schedule by providing technical assistance to the local compressor company.

The parent companies, Hitachi, Matsushita, Mitsubishi Electric and Toshiba of Japan, and the Japan Electrical Manufacturers Association (JEMA), launched a joint project to assist the local company by providing the technology necessary to improve reliability, and help with its implementation. This technical support was provided voluntarily, and the four companies were able to meet their phase-out schedule.

This phase-out date for household refrigerators, the end of 1996, was aggressive. It was only one year later than the phase-out date for developed countries, and it was the world's first phase-out in an entire industry subsector for any developing country. Dr. Stephen Andersen of the U.S. Environmental Protection Agency attended several follow-up meetings in Thailand and worked with the Thai compressor manufacturers' parent company in the United States to promote their participation in the technical support project. Because the Thai government realized that the ozone layer could only be protected if manufacturing shifted out of CFCs, they took measures to prohibit the manufacture or import of foreign CFC refrigerators after 1996. This quick action and support by the Thai government encouraged the refrigerator manufacturers to work hard to achieve the CFC phase-out. It is important to note an additional reason why this is a particularly inspiring success story. The Japanese companies which helped the Thai company develop and produce their HFC-134a compressors were their direct competitors which supplied the other three refrigerator manufacturers in Thailand.

The Vietnam Corporate Pledge

Vietnam is in the early stages of development. However, the use of ozone-depleting substances (ODS) will increase as industrialization increases, unless measures are taken to help Vietnam avoid investing in ODS-dependent technologies. It is very important for the global environment and for economic efficiency to stop ODS use in these early stages of economic development.

In August 1994, I visited Vietnam to discuss and promote ozone layer protection in Vietnam. I was accompanied by Dr. Stephen Andersen, Dr. Margaret Kerr of Nortel, Mr. Shinichi Ishida of Hitachi, and Dr. Viraj Vithoontien of the United Nations Environment Programme (UNEP). We visited Vietnam Motors Corporation, which was assembling vehicles using parts from Japan, Korea, and Germany.

Only two or three automobiles were being assembled in a very large factory, indicating they had plans for significant expansion. During our tour, we noticed that an automobile being assembled was equipped with an air conditioner using CFC-12. We immediately realized the importance of early action to avoid the rapid expansion in the use of CFCs in Vietnam.

The following year, in September 1995, we organized a conference entitled "Scientific Meeting for ODS Elimination" in Hanoi, Vietnam. Approximately twenty experts on cleaning, refrigerants and foam from Japan, the United States, the European Union, and UNEP participated. Eight members from the Technology and Economics Assessment Panel (TEAP) and the Solvent, Coating & Adhesives Technical Options Committee also participated.

The following pledge, which had been signed by leading multinational companies from around the world, was announced publicly at the conference:

> Our company pledges to invest only in modern, environmentally acceptable technology to avoid the use of chlorofluorocarbons (CFCs), halons, carbon tetrachloride, and 1,1,1 trichloroethane. Our company also pledges to limit the use of transitional substances such as hydrochlorofluorocarbons (HCFCs) when suitable replacements become available. We also encourage our joint venturers and suppliers to make this pledge.

Forty-three leading companies, twenty-five from Japan and eighteen from the U.S. and other countries, listed below, joined in the pledge. They are: Asahi Glass,

Asea Brown Boveri, AT&T, British Petroleum, British Petroleum Vietnam, Carrier, The Coca Cola Company, Daihatsu, DuPont, Ford, Fuji Electric, Fuji Heavy Industries, Hewlett-Packard, Hino, Hitachi, Honda, Honeywell, ICI, Isuzu, Kawasaki Heavy Industries, Lufthansa, Matsushita Electric, Mazda, Meidensha, 3M, Mitsubishi Electric, Mitsubishi Heavy Industries, Mitsubishi Motors, Motorola, Nissan, Nissan Diesel, Nortel (Northern Telecom), Sanyo, Seiko Epson, Sharp, Suzuki, Taiwan Fertilizer Company, Toshiba, Toyota, Trane, Yamaha, Yaskawa, Vulcan Materials and UNISYS.

Leadership Workshop for Climate Protection and Semiconductor Agreement

From April 2–4, 1997, the Leadership Workshop for Climate Protection was held in Tokyo with sponsorship by seven Japanese industrial associations: Electronic Industries Association of Japan (EIAJ), Federation of Electric Power Companies (FEPC), Japan Automobile Manufacturers Association (JAMA), JEMA, Japan Facility Management promotion Association (JFMA), Japan Industrial Conference on Ozone Layer Protection (JICOP) and Japan Refrigeration and Air Conditioning Industry Association (JRAIA).

The workshop addressed the reduction of global warming gases, but was patterned after our success in ozone layer protection. This workshop was defined as a "Pathfinder Meeting" and was successful in encouraging industry and governments to cooperate on climate protection initiatives. Ozone layer protection efforts were successful because of global leadership and cooperation. Climate protection, we believed, would only be successful by following the lessons we learned through implementing the Montreal Protocol. The conference covered a wide range of topics important to launching successful climate protection projects, including concept, strategy, policy, financing, and economics. In addition to these broad topics, the meeting identified issues and opportunities relevant to specific industries, such as automobiles, semiconductors, air conditioners, household refrigerators, chemical processes, electric utilities and energy efficiency.

Over the three days of the workshop, over 100 people attended from Japanese government, associations and private companies. The organizers requested that the semiconductor manufacturing process, be a priority for quick action to reduce perfluorocarbons (PFCs). The EIAJ announced the "Voluntary Action Plan on Reduction of PFC Emissions" in response to the Ministry of International Trade and Industry's request to the Association to develop an action plan to seek cost-effective reductions in the emissions of PFCs. These chemicals are of particular concern because of their significant global warming potential and extremely long atmospheric lifetimes. This Voluntary Action Plan is regarded as a comparable approach to the voluntary measures taken by the U.S. Environmental Protection Agency and a number of U.S. semiconductor manufacturers. The meeting participants welcomed the announcement and affirmed the need for international cooperation to ensure the successful implementation of the voluntary projects.

At the Fourth International Environment, Safety and Health Conference on the Semiconductor Industry held in June 1997 in Milan, Italy, Japanese, European, Korean, Taiwanese & American industry experts pledged cooperation on a number of issues including PFC emissions.

Conclusions

Based on these experiences, I believe there are three elements which are crucial to successfully achieving global environmental objectives:
1) global leadership and commitment to cooperation;
2) an early action plan from the industry; and
3) cooperation between government and industry.

Japan and the United States held a series of conferences to promote technology transfer in Southeast Asia. They were:
1) Bangkok, Thailand, in 1992;
2) Kuala Lumpur, Malaysia, in 1993;
3) Bangkok, Thailand, in 1994;
4) Jakarta, Indonesia, in 1994;
5) Bangkok, Thailand, in 1995;
6) Hanoi / Ho Chi Minh, Vietnam, in 1995;
7) Bangkok, Thailand, in 1996;
8) Jakarta, Indonesia, in 1996;
9) Manila, Philippines, in 1996.

In addition, we conducted the following series of Environmental Leadership Workshops:
1) Woods Hole, Massachusetts, in 1991;
2) Yountville, California, in 1992;
3) Osaka, Japan, in 1993;
4) Nara, Japan, in 1995; and
5) the Pathfinder Meeting in Tokyo, Japan, in April 1997.

Many people from Japan, the United States and the European Union were instrumental to the success of these workshops and projects presented here. Special recognition should go to the individuals and organizations who made these so successful, including UNEP/IE; UNEP Multilateral Fund; UNEP Technology and Economics Assessment Panel and Solvent, Coatings and Adhesives Technical Options Committee; the World Bank, the U.S. Environmental Protection Agency; the International Cooperative for Environmental Leadership (ICEL); the Japanese Ministry of International Trade and Industry (MITI), and the seven Japanese industrial associations who sponsored the Pathfinder meeting for climate change in April, 1997.

CHAMPIONS OF OZONE LAYER PROTECTION

Stephen O. Andersen

The Montreal Protocol is universally recognized as a scientific, diplomatic, regulatory, business, and technology cooperation success. It is a scientific success because early evidence was persuasive enough that governments took precautionary action. It is a diplomatic success because it demonstrated that countries can work together to resolve global environmental security threats. It is a regulatory success because the treaty was crafted to assure ozone protection but made flexible to allow each country to craft domestic approaches that are appropriate to national, industrial, and consumer situations. It is a business success because new technology was commercialized rapidly enough to replace, at affordable prices, the previous uses of ozone-depleting substances. Foremost, it is a technology cooperation success because industry put protection of the ozone layer above competitive concerns and generously shared non-proprietary information.

All these successes depended on the work of individuals and organizations I call "Champions of the Ozone Layer." There have been Champions in science, diplomacy, regulation, environmental advocacy, and industry. Three hundred and twenty Champions are recognized in the new Environmental Protection Agency (EPA) book: "*Champions of the World: Stratospheric Ozone Protection Awards*"; from the following twenty-five countries: Australia, Belgium, Brazil, Canada, Chile, France, Germany, India, Ireland, Japan, Kenya, Malaysia, Malta, Mexico, the Netherlands, Norway, Poland, Singapore, Sweden, Switzerland, Taiwan, Thailand, United Kingdom, United States, and Venezuela. My presentation today elaborates on military and industrial Champions who have earned the Environmental Protection Agency "Best-of-the-Best Stratospheric Protection Award" which will be presented Sunday night 14 September, 1997 at the Technology and Economics Assessment Panel (TEAP) dinner celebrating industry contributions to ozone layer protection.

Seventy-one individuals, corporations, and associations are the *crème de la crème* among those who have previously won EPA's Annual Stratospheric Ozone Protection Awards. The seventy-one recipients were nominated by their peers to be recognized for outstanding accomplishments in protecting the Earth's ozone layer. Winners are from Australia, Brazil, Canada, Germany, Japan, Malaysia, Mexico, the Netherlands, Switzerland, United Kingdom, United States, and Venezuela. Champions made extraordinary contributions in a variety of ways:

1) DuPont made an about face and announced that scientific evidence was conclusive that CFCs destroy the ozone layer. With industry support, political decisions were possible.
2) AT&T broke ranks with industry associations and announced a new solvent, made from oranges, that cleaned as effectively as CFC-113. Market forces encouraged commercialization of alternatives.
3) Nortel and Seiko Epson were first to set challenging phase-out goals far more stringent than the Montreal Protocol. Leadership inspired managers and engineers to innovate.
4) The Foodservice and Packaging Institute (FPI) with the support of the Environmental Protection Agency, Friends of the Earth, The Natural Resources Defense Council and the World Resources Institute announced that United States foodservice packaging companies would phase out CFC use by December 1988, the world's first voluntary national CFC phase-out. This demonstrated the opportunity for cooperation between governments, industry, and environmental non-governmental organizations (ENGOs).
5) The Japan Electrical Manufacturer's Association (JEMA) The Japan Industrial Conference on Ozone Layer Protection (JICOP), The Industry Cooperative for Ozone Layer Protection (ICOLP), and the Halon Alternatives Research Corporation (HARC) mobilized like-minded companies to cooperate to speed commercialization of non-proprietary technology. This empowered leadership companies to undertake market transformations.
6) The United States military surprised skeptics by acknowledging their responsibility as customers and spearheading technical innovation, green procurement, and market influence. Military leadership became contagious as the North Atlantic Treaty Organization took cooperative action to encourage the use of alternatives to CFCs, halons, and methyl chloroform.
7) Mexican environmental and industry leaders jointly declared that they preferred to proceed at the same pace as developed countries enabling them to leap-frog inferior and obsolete CFC equipment and use cutting-edge technology. Other developing countries soon joined Mexico in supporting more stringent controls under the Protocol.
8) Thailand went one step further. With the help of the Japan Ministry of International Trade and Industry, UNEP, JEMA, JICOP, ICOLP and U.S. EPA, it was announced that multinational leadership companies would phase out in Thailand within one year of their domestic phase-out schedule. By 1997, the first phase-out of CFCs used in the manufacture of refrigerators was accomplished, and Thailand became the first developing country in the world to

use trade controls to protect the global environment—prohibiting manufacture and import of CFC refrigerators.

This experience with the Montreal Protocol holds several lessons for the world as it confronts the problem of climatic change.

1) The controls originally introduced under the Montreal Protocol in 1987 motivated industry. Although little technology had yet been identified, industry moved quickly to commercialize alternatives and substitutes. The December 1997 Kyoto Protocol on Climate Protection may prove to be a similar precondition to gaining support from industry and to developing technology.

2) Corporate environmental goals focus company priorities and motivate suppliers. This is now beginning to repeat itself over global warming. British Petroleum has heeded the science and announced that they will act now to protect the climate by becoming a world leader in photovoltaic manufacturing. Dow has set a goal of improving energy efficiency by two percent a year and Mitsubishi Electric and Philips have announced targets of increasing it by twenty-five percent by 2010. General Motors has introduced an electric vehicle, Honda has an ultra low emission vehicle, Toyota is marketing a hybrid vehicle, and Mitsubishi has commercialized direct fuel injection. ENRON has developed a portfolio of climate friendly energy supply technology for its customers. Whirlpool has commercialized the world's most energy-efficient refrigerator and is advocating stringent regulation to motivate its competitors.

3) Partnerships and associations between industry and government attract companies taking leadership. The Industry Cooperative for Ozone Layer Protection has been reorganized as the International Cooperative for Environmental Leadership (ICEL) with a climate protection agenda. The organizers of the Alliance for Responsible Atmospheric Policy have formed the International Climate Change Partnership (ICCP) to guide climate negotiations towards flexible, performance-based regulations.

4) Information is critical. Industry experts on the Montreal Protocol's Assessment Panel catalogued and evaluated the best technologies. The UNEP Industry and Environment Programme office in Paris provided publications, databases, and on-line access to information, and organized networks and regional offices to guide the selection of technology. The Convention on Climate Change has now reached the stage where increased industry participation is needed and it is considering how UNEP can repeat its success in providing information that helped to protect the ozone layer.

5) Companies working together to commercialize alternatives to ODS soon discovered that the transition was far less difficult and much less expensive than most *ex ante* studies had predicted. In many cases, the CFC phase-out became the rallying point to completely rethink products and redesign manufacturing. This technical leap forward improved product performance and saved manufacturers and their customers money. Climate protection will be difficult, but may be far less daunting than currently predicted.

BEST-OF-THE-BEST STRATOSPHERIC OZONE PROTECTION AWARDS

Associations
- Air Conditioning and Refrigeration Institute (ARI)
- Alliance for Responsible Atmospheric Policy (ARAP)
- Australian Fluorocarbon Consumers and Manufacturers (AFCAM)
- Halons Alternatives Research Corporation (HARC)
- Institute for Interconnecting and Packaging Electronics Circuits (IPC)
- CFC Benchmarking Team
- International Cooperative for Ozone Layer Protection (ICOLP)/International Cooperative for Environmental Leadership (ICEL)
- Japan Electrical Manufacturers' Association (JEMA)
- Japan Industrial Conference for Ozone Layer Protection (JICOP)
- Mobile Air Conditioning Society (MACS)

Individuals
- Ward Atkinson
- James Baker
- Jay Baker
- Jonathan Banks
- Walter Brunner
- Suely Machado Carvalho
- David Catchpole
- David Chittick
- Jorge Corona de la Vega
- Philip J. DiNenno
- Stephen P. Evanoff
- Kevin Fay
- Joe Felty
- Arthur FitzGerald
- Yuichi Fujimoto
- Kaichi Hasegawa
- Andrea Hinwood
- Michael Jeffs
- Margaret G. Kerr
- Joel Krinsky
- Lambert Kuijpers
- Colin Lea
- Eduardo Lopez
- Mohinder Malik
- Melanie Miller
- John Minsker
- Mario Molina
- E. Thomas Morehouse
- David Mueller
- Tsuneya Nakamura
- Richard Nusbaum
- Simon Oulouhojian
- Jose I. Pons
- Sherwood Rowland
- Ronald W. Sibley
- Gary Taylor
- Daniel Verdonik
- Gary Vest
- Masaaki Yamabe
- Hideaki Yasukawa

Corporations & Military Organizations
- Asahi Glass
- The Coca-Cola Company
- DuPont
- Hitachi
- IBM
- ICI
- Lockheed Martin
- Lufthansa
- Malaysian Ministry of Science, Technology and the Environment
- Minebea Group Companies
- Mitsubishi Electric

- Nissan
- Nortel
- Raytheon TI Systems
- Seiko Epson
- Thiokol/NASA
- 3M Pharmaceuticals
- U.S. Air Force (Titan)
- U.S. Army Acquisition Pollution Prevention Support Office (AAPPSO)
- U.S. Department of Defense (DoD)
- U.S. Naval Research Laboratory
- U.S. Naval Surface Warfare Center

CLOSING COMMENTS

Gary Taylor and E. Thomas Morehouse

Gary Taylor

A balance of interests is critical to achievement of success. There is a natural tendency to try and ignore what you do not wish to hear by not involving those who seem to be fixed in their opposition. However, views not expressed within a group are often not addressed by the group and as a result those with opposing views are given the opportunity to position the process as unfair because there was inadequate response to their concerns.

All stakeholders must be offered the opportunity to participate; however, the price of entry must be ownership of the result.

The term industry representation is often used with little clear understanding—unfortunately it seems to often mean "not government" and not an environmental "NGO". The term is too broad to clearly identify important stakeholders from the private sector. Chemical producers, equipment manufacturers and users could be considered as broad categories within the "industry" sector. However, there are often major differences between the view and perceptions of what are problems and what are opportunities.

For some producers the goal was to simply produce any acceptable chemical alternatives and retain market position. However, for some producers their core business is a halogen and the ozone-depleting substance (ODS) is merely a way to market the core element. When that halogen is not used in replacements and alternatives, the Montreal Protocol poses a serious threat to the fundamental business of the company.

Some equipment manufacturers saw an opportunity to continue a business with an alternative or substitute that forced customers to retrofit existing equipment. Manufacturers make and sell equipment. The ODS or alternative is merely a material to wrap metal or plastic around. In some cases, users switched technologies and abandoned traditional suppliers.

On several Technical Options Committees (TOCs) "industry" was manufacturers and on several other TOCs "industry" was users. On one TOC, users dominate "industry" representation. Users recommended an early production phase-out based on use of recycled ODS for critical uses. Manufacturers were not in favour of early production phase-out because alternatives were not available. However, now that there are alternatives for many applications they now lobby strongly for early destruction of existing ODS and early retirement of equipment.

E. Thomas Morehouse

An important element of success to the Montreal Protocol mentioned by each of our panelists was something which is difficult to quantify, and even harder to predict. I am talking about the intangible will of the human spirit. Margaret Kerr made the connection between the seemingly disparate elements of her portfolio at Nortel—human resources, employee satisfaction, health and well-being, business ethics, security, customer satisfaction, and environmental affairs. These are the human dimensions to corporate success which we have not yet figured out how to quantify accurately on the balance sheet. Yuichi Fujimoto talked about inspired individuals from Japanese corporations who made difficult decisions to lend a hand to companies in developing countries. In giving Thai refrigerator factories the ability to use new refrigeration technology, Japanese companies also gave them the know-how to manufacture higher quality compressors. This is noteworthy because these companies voluntarily gave their competitor the ability to compete more effectively. This enabled Thailand to become the first Article 5(1) country to phase-out chlorofluorocarbons (CFCs) from domestic refrigerators without harm to its domestic industry. This remarkably altruistic act by multinational corporations was made possible because people in leadership positions made decisions to act according to conviction. Steve Andersen gave example after example of individuals who assumed similar leadership roles because they believed in the cause of ozone layer protection, felt strongly about their personal responsibility to care for the environment, and decided that their companies, government agencies or academic institutions should also act responsibly. He characterized them as the champions of the ozone layer.

In 1987, the science of ozone depletion would not have held up in a court of law; there was little confidence technologies could be found to replace CFCs, and the industries with the most financial interest in the CFC business opposed a treaty. Yet a modest treaty was negotiated, and champions emerged from each of the industrial sectors affected by the Montreal Protocol. As a result, a wide range of new technologies were born. Necessity really is the mother of invention, and throughout the saga of the Montreal Protocol, as science continued to provide the necessity, engineers, through their creativity and innovation, continued to provide the inventions. But in 1987, no one would have believed a complete phase-out possible and no one could have predicted the emergence of champions. However, as the science became more compelling and indicated a more urgent need to eliminate ozone-depleting chemicals, technology rose to the challenge, or more appropriately, individuals rose to the challenge and found the technological solutions. As the London and Copenhagen amendments further restricted consumption, the engineers found the technologies which enabled the restrictions agreed to on paper, to become a reality on factory floors, and in the products people use in their daily lives. Success inspired confidence, and confidence inspired yet more success.

Difficult decisions will be made this December in Kyoto regarding protection of the global climate, and there are some important lessons to be learned from the Montreal Protocol experience. One is not to underestimate or dismiss the will of the human spirit to rise to the challenge. Just as technologies delivered through creativity and innovation by committed and inspired individuals enabled the CFC phase-out, champions will undoubtedly emerge to provide the technological answers to climate protection. And just as these innovations which protect the ozone layer could not have been predicted in 1987, today we cannot predict the technologies which will be developed to protect the climate in the years ahead. This should not be an excuse for inaction, but a challenge to act according to conviction.

ANNEX 1

Ozone Protection Chronology

1840	- Discovery of ozone.
1879–81	- Ozone detected in the earth's stratosphere.
1928	- Invention of CFCs.
1957	- International Geophysical Year. - Establishment of Dobson stations.
1966	- John Hampson is first to argue that hydrogen compounds may be able to alter the ozone balance.
1970	- Paul Crutzen and Harold Johnston suggest that nitrogen compounds may be able to destroy stratospheric ozone.
1971	- U.S. Congress orders a report on the potential damage to the ozone layer by supersonic transport (SST).
1973	- Delegates at a conference in Kyoto discuss the suggestion that free chlorine may affect the ozone levels in the stratosphere (September).
1974	- Mario Molina and Sherwood Rowland publish their theory about the impact of CFCs on the ozone layer (June). - The U.S. Department of Transportation's Climate Impact Assessment Program releases a report arguing that a large fleet of SSTs would pose significant damage to the ozone layer.
1975	- The report of the U.S. Task Force on Inadvertent Modification of the Atmosphere finds that the Molina and Sherwood CFC theory is a "legitimate cause for concern" (June).
1976	- UNEP takes steps to convene an international meeting to consider the CFC-ozone link (March-April). - The U.S. National Academy of Sciences (NAS) releases its first report on the ozone layer, predicting most likely depletion to be between 6% and 7.5% (September).
1977	- UNEP's World Plan of Action on the regulation of CFCs adopted in Washington D.C. (April).

	- Ozone protection amendment to the U.S. Clean Air Act (August). - Creation of the UNEP-sponsored Coordinating Committee on the Ozone Layer (CCOL), Geneva (November). - S. C. Johnson first company to phase out CFCs in aerosols products.
1978	- The United States bans nonessential uses of CFCs, followed in later years by bans in Canada, Norway and Sweden (October).
1979	- A second U.S.-NAS report estimates eventual ozone depletion at 16.5 percent (November). - The Federal Republic of Germany hosts the second international conference on regulating CFCs (December).
1980	- The European Communities, having reduced aerosol use by thirty percent, enact a cap on capacity. - The UNEP Governing Council calls for cuts in the production of CFC-11 and CFC-12 (April).
1981	- The UNEP Governing Council calls for the development of a global framework convention for the protection of the ozone layer (May). - NASA scientist Donald Heath announces that satellite measurements show signs of ozone layer depletion (August).
1982	- First meeting of the Ad-hoc Working Group of Legal and Technical Experts for the Preparation of a Framework Convention, Stockholm (January). - A third NAS report predicts depletion at five to nine percent (March).
1983	- Norway, Sweden and Finland propose a worldwide ban on CFCs in aerosols, and limitations on all uses of CFCs (April).
1984	- A fourth NASA report reduces expected ozone depletion to between two and four percent (February). - Meeting of the Toronto Group, a coalition of like-minded nations committed to protecting the ozone layer (September).
1985	- Adoption of the Vienna Convention for the Protection of the Ozone Layer (March). - In a paper published in *Nature*, members of the British Antarctic Survey, led by Dr. Joe Farman, present evidence of a significant ozone "hole" over Antartica (May).

1986	- A NASA/WMO/UNEP Ozone Assessment Report finds evidence of reduced stratospheric ozone (January).
- The first of two workshops on economic issues related to the control of CFCs (part of a series of workshops designed to advance the drafting of a Protocol) is held in Rome (May).
- CFC manufacturers suggest that safe substitutes for CFCs might be possible, if the price were high enough (June).
- A second Economic Workshop is held in Leesburg, Virginia (September)
- The Alliance (an industry group) announces that it will support limits on CFC production (September).
- Ozone Trends Panel (OTP) formed (October).
- First session of the Ad Hoc Working Group of Legal and Technical Experts for the Elaboration of a Protocol on the Control of CFCs to the Vienna Convention for the Protection of the Ozone Layer, Geneva (December).
1987	- UNEP convenes scientists to discuss models of ozone depletion in Würzburg, West Germany (April).
- G-7 Venice Summit lists stratospheric ozone depletion first among environmental concerns (June).
- Adoption of the Montreal Protocol (September).
- Members of the Airborne Antarctic Ozone Experiment (part of the OTP) conclude that chlorine chemicals are the primary cause of ozone depletion in Antarctica (October).
1988	- OTP Report identifies ozone losses in the southern and northern hemispheres (March).
- DuPont announces that it will stop manufacturing CFCs as substitutes become available (March).
- The Vienna Convention enters into force (September).
- Ozone Depletion Conference in London (November).
1989	- The Montreal Protocol enters into force (January).
- London Conference on Saving the Ozone Layer (March).
- Helsinki Declaration adopted at the first Meeting of the Parties (MOP-I) of the Montreal Protocol (May).
- Global Atmosphere Watch (GAW) established.
- UNEP publishes a Synthesis Report, incorporating results of scientific, economic, environmental and technical assessments (November).
- UNEP forms Science, Effects, Economic and Technology Assessment Panels, along with Technical Options Committees.

1990	- London Amendments to the Montreal Protocol adopted at MOP-II (June). - Interim Multilateral Ozone Fund established (September).
1991	- MOP-III, Nairobi, Kenya (June). - Ozone action-programme established. - WMO and UNEP issue a report suggesting that ozone depletion has progressed more rapidly than predicted (October).
1992	- London Amendments enter into force (August). - Copenhagen Amendments to the Montreal Protocol adopted at MOP -IV, (November). - The Multilateral Fund is made permanent. - NATO writes to UNEP Executive Director supporting proposed London Amendment accelerating controls on ODS.
1993	- MOP-V, Bangkok, Thailand (November).
1994	- Copenhagen Amendments enter into force (June). - MOP-VI, Nairobi, Kenya (October).
1995	- Vienna Amendments to the Montreal Protocol adopted at MOP-VII, Vienna, Austria (December).
1996	- MOP-VIII, San José, Costa Rica (November).
1997	- MOP-IX, Montreal, Canada (September).
1998	- MOP-X, Cairo, Egypt (November).

ANNEX 2

Tenth Anniversary Colloquium

LESSONS FROM THE MONTREAL PROTOCOL

A statement of findings from the Colloquium, held on September 13, 1997 at Le Centre Sheraton Hotel, Montréal, Québec, Canada, issued on behalf of the International Advisory Committee and National Organizing Committee. Sponsored as part of the Tenth Anniversary Celebrations Program by Environment Canada, with the assistance of l'Université du Québec à Montréal and the Canadian Global Change Program of the Royal Society of Canada, and with the generous financial support of 3M and Nortel.

International Advisory Committee

- Pieter Aucamp, South Africa
- Suely Machado Carvalho, Brazil
- John Hollins, Canada
- Winfried Lang, Austria
- Nelson Sabogal, Colombia
- Jan C. van der Leun, the Netherlands

National Organizing Committee

- Angus Fergusson, Environment Canada
- Sonja Henneman, Environment Canada
- Claude Lefrançois, Royal Society of Canada, Canadian Global Change Program
- Philippe Le Prestre, Université, du Québec à Montréal
- Yarrow McConnell, Royal Society of Canada, Canadian Global Change Program
- Sara Melamed
- John D. Reid, Environment Canada
- Robert Saunders
- Hague Vaughan, Environment Canada

Introduction

On September 16, 1987, the Montreal Protocol on Substances that Deplete the Ozone Layer, an agreement under the 1985 Vienna Convention on the Protection of the Ozone Layer, was signed by twenty-four countries. Today, the number of Parties to the Protocol has grown to 162, accounting for more than ninety-five percent of the people of planet Earth.

Indeed, the tenth anniversary of the Montreal Protocol is a cause for global celebration. We have learned that international environmental agreements can work and are workable. For that achievement, the world community is to be congratulated. The Montreal Protocol underscores the universal recognition that nations are interdependent; that the ozone layer cannot be restored and protected without the active participation of all the countries of the world—developing and developed.

The world was up to the job

On this anniversary, we celebrate the process of reaching and strengthening an important global commitment along with the building of trust among governments, industry and communities; a commitment not driven by narrow economic or political interests but rather through consensus on the need to act.

The success of the Protocol has been acknowledged even by skeptics of the effectiveness of international environmental agreements. The Worldwatch Institute, in its State of the World 1997 Report, notes, "At a time when progress in reversing other global environmental trends is distressingly slow, it is reassuring to see that in responding to the threat of ozone depletion, the world was up to the job." Now we must continue our work.

The movement to renew the commitments of the Montreal Protocol, in the spirit of 1987, will help establish the necessary guidelines for continuing global cooperation in protecting the ozone layer.

Background

At the Tenth Anniversary Colloquium held on September 13, 1997, the participants reviewed the important progress achieved under the Montreal Protocol from several perspectives and pondered the lessons learned from ten years of experience.

The Colloquium examined the central roles played by the natural and social sciences, policy, and technology as well as science and technology assessment in the development and implementation of the Protocol. The following is presented to the meeting of the Parties to the Protocol on behalf of the National Organizing Committee and International Advisory Committee of the Colloquium.

Overview

It is clear that the public's attention has been seized by the ozone depletion issue, driven by health concerns for UV-radiation exposure, better awareness and knowledge about radiation and its health effects, compelling images of the deepening Antarctic ozone hole and effective media liaison by advocacy groups. Indeed, public support for action combined with the efforts of governments, diplomats, industry, scientists and concerned citizens worldwide, have been key to the achievements of the Protocol.

Hailed as a model for other international environmental agreements, the Protocol owes its success to the collegial and cooperative approach it worked to establish, especially between developed and developing countries. This was a highly effective way to tackle the problem. Its targeted, non-punitive, apolitical and science-based approach in working with the biggest consumers of ozone-depleting substances, i.e., industrialized countries, has also borne fruit in achieving reduction objectives over the last decade. There could be no better example of the power of "Working Together".

Through the work of the Protocol, scientific consensus on the dangers of ozone depletion was achieved and perhaps for the first time diplomats listened to the advice of scientists as they explained the real threat facing the planet. Through the work of the Protocol, industrialized nations and developing countries forged an innovative partnership in a global effort Through the work of the Protocol, the world has received new awareness, new science, new technology and a new form of cooperation.

Significant challenges remain as the Protocol launches into its second decade, notably harmonization of commitments, preventing illegal production and smuggling, and continuing to advance control provisions for ozone-depleting substances. The objective must be to reinvigorate commitment of all parties to the Protocol and its guiding principle: "The delicate balance supporting human life on earth depends on the ozone layer."

> The main lesson from a decade of experience with the Montreal Protocol is that establishing the reality of the threat is a prerequisite to purposeful international action. Once the threat is acknowledged, governments and societies have demonstrated their willingness to accept the costs associated with informed action provided the latter does not jeopardize major established social, medical, economic or technological benefits. Industry is ingenious in finding alternative ways of delivering these benefits. In the final analysis, sound information, effective institutions, a spirit of cooperation, and inspired individuals, all contribute to bridging political differences in favour of the pursuit of the common interest.

Natural sciences: What we have learned

Vital to the success of the Protocol, the scientific contribution to the issue arose from a broad base of academic research, the quest to better understand our world, and mission-oriented science to address problems. The science underlying the Protocol evolved gradually; from research on the structure of the atmosphere, the spectrum of solar radiation and its effects on organisms; from studies to improve weather forecasting, and from environmental concerns over emissions from supersonic transport aircraft. Maintaining a base of research remains imperative.

Atmospheric monitoring has been essential to understanding the reality of the threat and its evolution—the accumulation of ozone-depleting substances in the atmosphere, and their effect on the ozone layer. With such key data, the need for action to stop the build-up of these compounds could be appreciated along with the realization that continual monitoring of the atmosphere is essential. It was also realized that while progress has been made, (in the first eight years following the signature of the Protocol, worldwide consumption of ozone-depleting substances dropped by seventy-three percent) there is still much to be achieved. If all the countries of the world were to meet their obligations, the ozone layer would only fully recover by the year 2050 at the earliest.

An improved understanding of ultraviolet radiation has provided the catalyst for actively pursuing the aims of the Protocol, and in the case of human health, has helped educate people to the hazard of sun exposure. A lack of good understanding of the effects of increased UV radiation on ecosystem and human health remains a concern.

The Protocol acted as a stimulus to new insights in the ozone layer science and improved understanding useful in dealing with the threat of climate variability and change. This is yet another example of how progress in one area helps build the knowledge base for another issue.

The lesson from the past ten years is the lesson of history; that environmental science and knowledge are an essential investment, much less expensive than ignorance.

Social sciences: The success of the Protocol policy

If the Montreal Protocol serves as a model for other international environmental agreements, it is most likely due to the realistic way it addressed the issue and the cooperative spirit it engendered.

From the outset, it was understood that the pressure on other resources, availability of clean water, and adequate food, tend to push global issues such as ozone depletion down on the national priority list of developing countries, just as pressures of national deficit, debt, and unemployment have in developed countries. So the Protocol established a partnership based on the principle of common but differentiated responsibility, recognizing that the circumstances of the low consumption countries, operating under Article 5, are different from those of the developed world. And since the problem originated in the industrialized world, it was there that action was initially taken, while giving developing countries a grace period. Further differentiation in the implementation was made for countries with economies in transition.

The most important feature of the Protocol was the innovative, dynamic and flexible arrangements that it put in place. The crafters of the new ozone environmental regime designed it in order to facilitate the integration of science into policy thereby allowing for adjusting phase-out schedules and controlling all ozone-depleting substances, not just those initially identified in the Protocol. They facilitated its implementation not only through the promotion of new principles of international cooperation, such as that of common but differentiated responsibility, but also through the creation of new institutions, such as the Implementation Committee and the Multilateral Fund. The innovative approach adopted by the Montreal Protocol in the field of rule-making and rule-implementation constitutes a major contribution to the development of international environmental law. Thus, ten years later, the pioneers of 1987 have bequeathed the international community a unique international instrument whose effectiveness continuously improves as scientific knowledge mandates and the political context allows. Subsequent agreements continue to benefit from these innovations.

Financial and technological transfers, along with the potential of trade restrictions, have proven a significant stimulus to action. Other necessary measures comprised facilitating and sometimes funding the establishment of adequate domestic environmental legislation, cost-effective administrative systems, and other incremental costs of adoption of technology. A targeted, non-punitive approach has assisted in cases of non-compliance.

Of particular success was the bilateral cooperation generated by the Montreal Protocol Multilateral Fund, established in 1990 with the aim of providing financial and technical assistance so that developing countries could meet their treaty obligations at no net cost to their economies.

Since the Fund was established, it has allocated over $1 billion for more than one thousand activities in 102 Article 5 countries. It is clear that the funding commitments from industrialized countries must be maintained if the gains made are not to be jeopardized. The work of the World Bank, UNDP, UNIDO and UNEP have all been significant, together with some highly effective bilateral cooperation.

The next decade will be a crucial period for the Protocol. Obligations for real reductions in developing country use of ozone-depleting substances reach their deadlines, testing their resolve and, importantly, that of developed countries to meet legitimate financial and technical cooperation requirements.

Technology: The need to innovate

The Protocol has had a marked impact on technology, specifically where it concerns the ban on CFCs, spawning a high level of innovation and business opportunity. Just as the electronic highway has revolutionized communications and forced the world to adapt accordingly, so too has the Protocol in stimulating the development of technology alternatives to ozone-depleting chemicals for uses such as refrigeration, foams, solvents, metal cleaning, dry cleaning, fire protection and aerosols. Business, industry, governments and consumers all recognize the need to evolve and innovate.

In accepting the ozone challenge, corporations can enhance their marketing strategies and respond more directly to consumer awareness by being the first in the marketplace to introduce safe products, so benefiting from a better public image and gaining a competitive edge.

Most of the initial cost estimates for technology substitutes proved too high. Industry costs were lowered when the creativity of engineers and business people was directed at the issue, and through tax incentives for research and development along with industry sector cooperation in pre-competitive research.

Here again, the Protocol relied on a balanced approach, recognizing that certain products, such as the metered dose inhaler (used to treat asthma), should be designated an "essential use exemption", and not be banned until appropriate alternatives were found.

There are now practical technology alternatives for virtually all uses of ozone-depleting substances. The challenge now, still not to be underestimated, is largely one of implementation in developing countries.

The importance of assessments: keeping politics at arm's length

There is much that is unique about the Montreal Protocol and the way it was developed. As research findings were so vital to the decisions leading to the agreement, scientific, environmental, technical and economic assessments are stipulated in the Protocol at least every four years. These assessments, entrenched in the Protocol in a way which puts them at arm's length from political considerations, provide a basis for further decisions on ozone-depleting substances and the actions necessary on an international basis.

It is important that assessments not be beholden to special economic or political interests. This is achieved by drawing on the top subject experts, wherever they may be found, in every aspect of the issue, and an exhaustive process of peer review. This impartial approach explains why there is generally no argument over the technical data presented.

The Protocol has successfully used the consensus arrived at on science and technology through the assessment process as a powerful driver to update the Protocol and its implementation. While considerable progress has been made, communicating assessment results so that they are understood by the whole range of stakeholders remains a challenge.

Index

A

Ad Hoc Solvents Working Group · 184
Adaptation · 2, 5, 7, 9, 19, 127, 130, 131, 133, 138
Albritton, Daniel · 6, 7, 24
Anderson, Steve · 24
Article 5 (1) countries · *See* Developing countries
Article 5 (1) status · 154, 155, 157, 158, 159, 160, 161
Assessments · 7, 84, 85, 127, 129, 133, 143
    scientific · 6, 8, 15, 21, 24, 29, 40, 48, 54, 67, 68, 69, 70, 71, 72, 73, 74, 75, 76, 77, 130 *See also* Technology and Economic Assessment Panel (TEAP)
Atmospheric Ozone Report · 128

B

Bakken, Per · 24
Basel Convention · 8, 23, 24, 81, 93, 96, 99, 100, 103
Benedick, Richard E. · 3, 8, 24, 127
Boshkov, Rumen · 24
Brinkhorst, Laurens Jan · 24
British Antarctic Survey · 77, 128, 218
Buxton, Victor · 3, 8, 24

C

Canada
    Atmospheric Environment Service · 59
    Environment Canada · 47, 48, 59, 62, 65, 77, 173
Cancer · 114
    Canadian Cancer Society · 63
    Frederick Cancer Research Centre · 50
    Skin cancer · 22, 48, 49, 53, 54, 55, 57, 58, 59, 63, 69, 120, 138, 173, 175, 176, 193
Carbon tetrachloride · 154, 175, 188, 203
Cataract · 22, 50, 55, 57, 120, 175, 176, 193
CFC
    and climate change · 76
    and ozone depletion · 14, 15, 22, 48, 69, 119
    and Vienna Convention · 82
    CFC-11 · 16
    CFC-113 · 86, 164, 184, 197, 199, 208
    CFC-12 · 16, 203
    concentrations · 2, 3, 16
    consumption/production · 2, 5, 14, 87, 101, 128, 158, 170, 172, 179, 180, 182, 183, 186,
    control costs · 183, 184, 185, 186
    emissions · 128
    invention · 14
    phase-out · 21, 23, 69, 88, 100, 102, 109, 128, 154, 158, 170, 177, 179, 180, 181, 183, 184, 187
    properties · 14, 48, 72
    substitutes · 17, 25, 69, 71, 86, 100, 103, 124, 175, 180, 186, 187, 202
    trade · 101, 103
China · 2, 4, 21, 88, 109, 118, 136, 199
Civil society · 3, 22, 24, 59, 60, 63, 65, 107, 119, 120, 137, 150, 187, 197
Claussen, Eileen · 24
Climate Change
    Clinton Administration · 138
    compared to ozone issue · 6, 7, 73, 74, 75, 76, 83, 87, 93, 95, 104, 105, 106, 108, 118, 119, 121, 125, 132, 135, 136, 138, 188, 209
    Framework Convention · 82, 96, 110
    greenhouse gas reduction · 187

Intergovernmental Panel on Climate Change (IPCC) · 70, 71, 75, 137
Kyoto Protocol · 73, 95, 103, 105, 137, 209
trade measures · 99
Common but differentiated responsibilities · 23, 105, 109, 177, 224, 225
Compliance · 5, 119, 133, 194
and other Multilateral Environmental Agreements · 93, 94, 96, 97n
costs · 181, 182, 185, 186, 188
Countries with Economies in Transition (CEIT) · 4, 153, 154, 155, 157, 158, 160
incentives · 4, 99, 103, 108, 109, 110n
monitoring · 85, 91
Montreal Protocol model · 4, 5, 91, 92, 93, 96n, 107
requirements · 6, 92, 95
Consensus · 5, 7, 8, 20, 21, 75, 87, 105, 113, 115, 123, 124, 125, 130, 150, 152, 167, 168, 169, 170
Convention on Biological Diversity · 82, 87, 88, 96
Convention on International Trade in Endangered Species of Wild Fauna and Flora (CITES) · 8, 99, 100, 103, 108, 111
Convention to Combat Desertification · 8, 23, 82, 83, 93, 96, 107, 110
Copenhagen Amendments · 161, 215
Cost-benefit analysis · 173, 174, 175, 177
Countries with Economies in Transition (CEIT) · 6, 84, 85, 88, 92, 109, 153, 154
and TEAP · 154, 155
Compliance/non-compliance · 4, 153, 154, 155, 160
Data reporting · 153, 159
Equity issues · 157
List and status · 161, 162
Crutzen, Paul · 14, 76, 217

D

Data reporting · 6, 20, 84, 91, 93, 156, 160
Developed countries · 137, 159, 180, 181, 193, 194, 201, 208
and climate change · 187
and environmental negotiations · 24
and Multilateral Fund · 115
and ODS consumption/production · 23, 170
and ODS phase-out · 7, 154, 204
and Technology and Economic Assessment Panel (TEAP) · 146
support for developing countries · 105, 113, 114, 121
trade · 188
Developing countries · 109, 121, 137, 157, 188, 214
and climate change · 95
and Global Environment Facility (GEF) funding · 110n
and industry · 102, 191, 192, 193, 194, 199
and Multilateral Environmental Agreements negotiations · 8, 24, 124
and Multilateral Fund · 20, 87, 102, 105, 108, 114, 115, 116, 177n, 225
and ODS consumption/production · 2, 17, 191, 192, 194
and ODS phase-out · 5, 9, 23, 87, 102, 146, 193, 194, 202, 208, 214
and Technical Options Committees (TOCs) · 149
and Technology and Economic Assessment Panel (TEAP) · 146, 165
incentives for · 23, 101, 105, 108, 109, 113, 114, 146
nutrition · 41
obligations · 9, 23, 105
Diplomacy · 13, 21, 74, 81, 83, 88, 99, 121, 127, 136, 207
DNA · 41, 43, 44, 45, 47, 52, 56
Dowdeswell, Elizabeth · 88
DuPont · 86, 186, 204, 208, 210

E

Eastern Europe · *See* Countries with Economies in Transition (CEIT)
Economic Commission for Europe (U.N.) · 107, 136
Equity issues · 4, 8, 9, 23, 88, 103, 104, 106, 109, 118, 120, 157, 213

Index   229

European Community · 102, 106, 124, 200, 201, 203, 205

F

France · 64, 136, 172, 207

G

General Agreement on Tariffs and Trade (GATT) · 105, 106, 108
Global commons · 135, 139
Global Environment Facility (GEF) · 109, 110, 115, 156, 157, 158, 161
Gorsuch, Anne · 86
Governance · 82
Greenpeace · 88
Group of 77 · 24, 124

H

Halons · 2, 7, 17, 20, 24, 72, 86, 100, 101, 102, 154, 175, 180, 181, 194, 203, 208, 210
  Technical Options Committee · 5, 144, 171, 172
HCFC · 17, 73, 101, 130, 139, 154, 175, 186, 203
  HCFC-22 · 184
HFC · 73, 104, 186
  HFC-134a · 187, 202, 203
Hodel, Donald · 138
Hydrocarbons · 175, 184, 193

I

Illegal traffic · 4, 9, 25, 93, 101, 103, 179, 187
Implementation Committee · 4, 85, 88, 92, 109, 156, 162
India · 2, 4, 21, 24, 88, 109, 111, 172, 199, 207
Industrialized countries · *See* Developed countries

Industry · 213
  adaptation · 102, 132, 164, 172, 180, 181, 184, 185, 187, 197, 202, 209
  and developed countries · 193, 199
  and developing countries · 102, 191, 192, 193, 194, 199, 202, 203, 204, 205, 206, 214
  and trade measures · 102
  co-operation · 83, 84, 87, 125, 145, 167, 176, 199, 204, 205, 207
  innovation · 132, 184, 185, 208
  opposition · 7, 86, 128, 168, 179
  role of · 24, 68, 73, 143, 144, 146, 163, 187, 191, 193, 196, 208, 214
Industry Cooperative for Ozone Layer Protection (ICOLP) · 184, 192, 208, 210
Irradiance · 29, 30, 31, 32, 33, 34, 35, 36, 52, 60

J

Japan · 6, 34, 45, 88, 207, 208
  and industries · 203, 205, 214
  and Thailand · 201, 203, 208
  and Vietnam · 192

K

Korea (Republic of) · 102, 203, 205
Kuijpers, Lambert · 6, 24, 153, 167, 210

L

Lang, Winfried · 8, 22, 24, 81, 96, 108, 111
Lebabty, Steve · 24
Leesburg Meeting (Virginia) · 83
Limited Test-Ban Treaty · 136
Lippmann, Walter · 74
London Amendments · 109, 115, 153, 156, 157, 161, 162
Long-range transboundary air pollution treaty (LRTAP) · 136

M

Market mechanisms · 8, 82, 86, 119, 132, 180, 181, 184, 185, 188, 208
Mateos, Juan Antonio · 24, 113, 117, 123, 125
McDonald, James E. · 57
Meeting of the Parties (MOP) · 1, 2, 5, 20, 85, 91, 143, 149, 154, 159, 161, 162, 169, 173
Metered-dose inhalers (MDI) · 5, 170
Methyl bromide · 2, 4, 9, 17, 21, 69, 100, 130, 132, 139, 151, 154, 169, 175, 179, 192, 194
  Technical Options Committee · 7, 144, 168, 169, 194
Methyl chloroform · 16, 21, 154, 175, 188, 208
Mexico · 21, 24, 88, 102, 199, 207, 208
Molina, Mario · 2, 13, 19, 22, 68, 76, 114, 210
Multilateral Fund · 4, 84, 115
  and Countries with Economies in Transition (CEIT) · 156, 157, 158, 159, 160, 161, 162
  and developing countries · 20, 87, 102, 105, 108, 114, 115, 116, 177n, 225
  creation · 20, 21, 24, 114, 125
  impact · 23, 85, 87, 104, 109
  implementing agencies · 85, 109, 110
  structure · 114

N

Nath, Kamal · 24
Nesbaum, Richard · 165
New Zealand · 21, 124
NGO · 8, 22, 82, 83, 85, 87, 88, 124, 125, 193, 208, 213
No-clean technology · 164, 165
Non-compliance · 4, 85, 91, 96, 97, 109, 160, 161, 162
  procedures · 5, 20, 21, 23, 84, 88, 93, 94, 108, 109
Nordic countries · 21, 82, 124

Northern Telecom (Nortel) · 88, 184, 193, 195, 196, 198, 199, 203, 204, 208, 211, 214
North-South · 23, 88, 113, 114, 115, 124

O

Obasi, Patrick · 88
Ozone hole · 3, 15, 16, 22, 35, 36, 59, 69, 71, 72, 77, 119, 128, 179
Ozone Secretariat · 85, 155, 156, 162

P

Partnership · 123, 124, 184, 209
Patten, Chris · 24
Perfluorocarbon (PFC) · 72, 104, 204, 205
  Persistent organic pollutants · 9, 24, 103,

R

Regime · 9, 109, 124, 135, 139
  acid rain · 136
  adaptation · 5–6, 24, 127, 130, 131, 133
  climate change · 70, 105, 108, 137
  compliance · 91–95, 96n
  definition · 108, 117
  design · 4, 117, 120–121
  interactions · 7, 106, 119, 124
  Montreal regime · 6, 8, 24, 108, 118, 125, 136
  structure · 3, 108
Reily, William · 24
Reports (Annual) · 92, 151, 165
Rio · *See* United Nations Conference on Environment and Development (UNCED)
Ristimaki, Illka · 24
Rome Meeting · 76, 124
Rowland, Sherwood · 14, 210, 217

Rummel-Bulska, Iwona · 22
Russia (Federation of) · 4, 6, 20, 81, 88, 92, 96, 124, 135, 153, 154, 155, 156, 158, 160, 162

S

Science
  and policy-making · 67, 70, 74, 76, 121, 146, 177, 225
  scientific uncertainty · 82, 104, 177, 192
  disagreements on · 6, 137
  role of · 1, 2, 3, 17, 21, 68, 75, 222
Scientists
  role of · 2, 3, 75, 83
Shultz, George · 81
Soviet Union · *See* Russia (Federation of)
Stockholm Conference · 23, 120, 125106
Széll, Patrick · 4, 8, 24, 107, 108

T

Technical Options Committees · 5, 7, 85, 144, 145, 149, 150, 151, 164, 165, 166, 167, 168, 169, 170, 172, 214
  Aerosols and Medical Sterilants · 165
  Halons · 5, 144
  Methyl Bromide · 7, 168, 169, 171
  Refrigeration · 144
  Solvents, Coatings and Adhesives · 144, 203, 205
Technology and Economic Assessment Panel (TEAP) · 5, 6, 7, 144, 145, 146, 149, 150, 163, 164, 170, 171, 173, 191, 203, 207
  and CEIT compliance · 154, 160
  and industry · 146, 150, 164, 168
  and international co-operation · 163
  and scientific objectivity · 167, 168
  history of · 129, 143, 144
  membership · 146, 149
  operation · 130, 144, 145, 150, 165, 166

  reports · 151, 158, 165, 168
  structure · 6, 85, 144, 150
  terms of reference · 149
Technology transfer · 9, 23, 24, 87, 88, 103, 105, 106, 146, 175, 177, 205
Thailand · 5, 88, 201, 202, 205, 207, 208, 220
Thatcher, Margaret · 88, 170
Tolba, Mostafa K. · 3, 6, 7, 8, 84, 115, 117, 123, 125, 127
Toronto Group · 128, 218
Trade
  General Agreement on Tariffs and Trade · 105, 106, 108
  World Trade Organization · 7, 105, 106, 108
  barriers · 7, 108
  illegal · 93, 96, 103, 110, 179
  measures · 3, 9, 84, 99, 101, 102, 103, 104, 105, 106, 108, 209
  regime · 7, 124
  restrictions · 7, 9, 84, 86, 99, 100, 102, 103, 104, 106, 108

U

United Kingdom (U.K.) · 24, 96, 135, 199, 207, 208
United Nations Conference on Environment and Development (UNCED) · 88, 110, 119, 120, 125
United Nations Development Programme (UNDP) · 85, 115
United Nations Environment Programme (UNEP) · 20, 102, 105
  and CEIT · 160
  and MLF · 85, 115
  and other MEAs · 23
  and ozone negotiations · 3, 22, 82, 114, 115
  and TEAP · 151
  assessment panels and reports · 39, 40, 67
  Industry and Environment Office · 87, 209
  ozone data · 128, 194, 209

United Nations Industrial Development Organization (UNIDO) · 85, 225
United States (U.S.)
  and climate change · 8, 24, 137, 138–139, 187, 188
  and illegal trade · 103
  and ODS phase-outs · 179, 181, 183, 185, 186
  and ozone negotiations · 3, 21, 82, 128
  and UV Index · 64
  Department of Energy (DoE) · 185
  Environmental Protection Agency (EPA) · 88, 185, 188, 192, 202, 204, 205, 208
  internal conflicts · 8, 124, 128, 131, 179
  cooperation with Thailand · 201
  cooperation with Vietnam · 192
  Military · 180, 184, 208
  ODS production and consumption · 179, 182, 183, 188
  ozone science research · 3, 48, 124
  trade sanctions · 102
  Reagan Administration · 81, 86, 88, 138
Usher, Peter · 22
UV Index · 3, 31, 34, 35, 60, 62, 63, 64, 65

World Meteorological Organization (WMO) · 3, 65, 67, 68, 88, 128, 218, 220
World Trade Organization (WTO) · 7, 105, 106, 108, 117

V

van der Leun, Jan C. · 24, 48, 49, 69, 221
Vienna
  Convention · 1, 2, 20, 22, 82, 88, 123, 131, 155, 162, 218, 219
  Meeting of the Parties · 2, 69, 96, 101, 154, 157, 169
Vietnam · 88, 192, 194, 199, 203, 205

W

Watson, Robert · 24
World Bank · 85, 109, 110, 115, 120, 199, 201, 205
World Health Organization (WHO) · 55, 65